Frontiers in Magnetic Materials

The book aims to provide comprehensive and practical guidance on magnetism and magnetic materials. It involves four parts, focusing on fundamental magnetism, hard magnetic materials, soft magnetic materials and other functional magnetic materials.

Part I highlights the ubiquity of magnetism and the close relationships between magnetic materials and our daily life. Perspectives on magnetism from Engineering and Physics are provided to introduce the two unit systems, followed by the origin and categories of magnetisms. An introduction of important parameters during magnetization and magnetic measurement techniques are then provided to lay a solid foundation for the readers for better understandings of the design and development of different magnetic materials.

Important magnetic materials are then introduced in the subsequent parts, delivering an overview of design principles, production technologies, research developments and real-world applications. For instance, rare-earth-free and rare-earth-based hard magnetic materials as well as soft magnetic materials such as Fe-based alloys, composites and ferrites are discussed. Other functional magnetic materials span a wide range, involving smart materials with magneto-X effects, together with magnetic materials for applications including electromagnetic wave absorption, biomedicine and catalysis, etc. For these magnetic materials, more emphasis is placed on the latest advances and interdisciplinary perspectives.

Frontiers in Magnetic Materials

From Principles to Material Design and Practical Applications

Chen Wu and Jiaying Jin

CRC Press
Taylor & Francis Group
Boca Raton London New York

CRC Press is an imprint of the
Taylor & Francis Group, an **informa** business

First edition published 2023
by CRC Press
6000 Broken Sound Parkway NW, Suite 300, Boca Raton, FL 33487-2742

and by CRC Press
4 Park Square, Milton Park, Abingdon, Oxon, OX14 4RN

CRC Press is an imprint of Taylor & Francis Group, LLC

© 2023 Chen Wu and Jiaying Jin

Reasonable efforts have been made to publish reliable data and information, but the author and publisher cannot assume responsibility for the validity of all materials or the consequences of their use. The authors and publishers have attempted to trace the copyright holders of all material reproduced in this publication and apologize to copyright holders if permission to publish in this form has not been obtained. If any copyright material has not been acknowledged please write and let us know so we may rectify in any future reprint.

Except as permitted under U.S. Copyright Law, no part of this book may be reprinted, reproduced, transmitted, or utilized in any form by any electronic, mechanical, or other means, now known or hereafter invented, including photocopying, microfilming, and recording, or in any information storage or retrieval system, without written permission from the publishers.

For permission to photocopy or use material electronically from this work, access www.copyright.com or contact the Copyright Clearance Center, Inc. (CCC), 222 Rosewood Drive, Danvers, MA 01923, 978-750-8400. For works that are not available on CCC please contact mpkbookspermissions@tandf.co.uk

Trademark notice: Product or corporate names may be trademarks or registered trademarks and are used only for identification and explanation without intent to infringe.

Library of Congress Cataloging-in-Publication Data
Names: Wu, Chen, 1985- author. | Jin, Jiaying, 1988- author.
Title: Frontiers in magnetic materials : from principles to material design and practical applications.
Description: First edition. | Boca Paton, FL : CRC Press, [2023] | Includes bibliographical references. |
Identifiers: LCCN 2022004999 (print) | LCCN 2022005000 (ebook) | ISBN 9781032106359 (hbk) | ISBN 9781032106410 (pbk) | ISBN 9781003216346 (ebk)
Subjects: LCSH: Magnetic materials.
Classification: LCC TK454.4.M3 W83 2023 (print) | LCC TK454.4.M3 (ebook) | DDC 621.34--dc23/eng/20220407
LC record available at https://lccn.loc.gov/2022004999
LC ebook record available at https://lccn.loc.gov/2022005000

ISBN: 978-1-032-10635-9 (hbk)
ISBN: 978-1-032-10641-0 (pbk)
ISBN: 978-1-003-21634-6 (ebk)

DOI: 10.1201/9781003216346

Typeset in Minion Pro
by SPi Technologies India Pvt Ltd (Straive)

Contents

Author Bio, xv

PART I **Fundamental Magnetism**

CHAPTER 1 ▪ Introduction to Magnetics   3
  1.1 INTRODUCTION TO MAGNETISM AND MAGNETIC MATERIALS   3
  1.2 VIEWS ON MAGNETISM AND TWO UNIT SYSTEMS   5
  REFERENCES   7
  FURTHER READING   7

CHAPTER 2 ▪ Origin and Categories of Magnetisms   9
  2.1 ATOMIC ORIGIN OF MAGNETISMS   9
      2.1.1 Arrangement of Electrons   9
      2.1.2 Electron Orbital and Spin Moment   12
      2.1.3 Total Magnetic Moment of the Atom   13
  2.2 FIVE TYPES OF MAGNETISMS   16
      2.2.1 Diamagnetism   16
      2.2.2 Paramagnetism   17
      2.2.3 Antiferromagnetism   20
      2.2.4 Ferromagnetism   21
      2.2.5 Ferrimagnetism   23
  REFERENCES   26
  FURTHER READING   27

Chapter 3 ▪ Important Parameters and Magnetic Measurements   29

3.1	BASICS OF MAGNETIZATION AND MAGNETIC PARAMETERS	29
3.2	MAGNETIC MEASUREMENT TECHNIQUES	33
	3.2.1 Measurement of Magnetization	33
	3.2.2 Measurements Based on Magnetic-X Effects	36
	3.2.2.1 Measurements Based on Magneto-Optical Effect	36
	3.2.2.2 Measurements Based on Magneto-Electric Effect	38
	3.2.2.3 Measurements Based on Magneto-Force Effect	38
	3.2.2.4 Measurements Based on Magnetic Resonance	40
	REFERENCES	41
	FURTHER READING	41

Part II   Hard Magnetic Materials

Chapter 4 ▪ Introduction to Hard Magnetic Materials   45

4.1	PERFORMANCE REQUIREMENTS FOR HARD MAGNETIC MATERIALS	45
	4.1.1 Intrinsic Magnetic Properties	45
	4.1.2 Hysteresis Loop	46
	4.1.3 Remanence	47
	4.1.4 Coercivity	49
	4.1.5 Energy Product	50
	4.1.6 Temperature Coefficients	51
4.2	DEVELOPMENT OF HARD MAGNETIC MATERIALS	52
4.3	APPLICATIONS OF HARD MAGNETIC MATERIALS	54
	4.3.1 Hard Magnetic Materials for Wind Power Generation	55

	4.3.2	Hard Magnetic Materials for New Energy Vehicle	55
	4.3.3	Hard Magnetic Materials for Information Technology	56
	4.3.4	Hard Magnetic Materials for Industrial Robot	56
REFERENCES			57
FURTHER READING			59

CHAPTER 5 ▪ Rare-Earth-Free Hard Magnetic Materials　　61

5.1	HARD FERRITES		61
	5.1.1	Introduction	61
	5.1.2	Crystal Structure and Magnetic Properties	62
	5.1.3	Synthesis Techniques	64
		5.1.3.1 Standard Ceramic Method	64
		5.1.3.2 Sol-Gel Method	66
		5.1.3.3 Coprecipitation Method	67
		5.1.3.4 Hydrothermal Synthesis	67
		5.1.3.5 Other Methods	68
	5.1.4	Substituted M-Type Ferrites	68
	5.1.5	Prospects and Future Challenges	70
5.2	ALNICO		71
	5.2.1	Introduction	71
	5.2.2	Timeline of Alnico Development	71
	5.2.3	Prospects and Future Challenges	76
REFERENCES			76
FURTHER READING			82

CHAPTER 6 ▪ Rare-Earth-Based Hard Magnetic Materials: SmCo　　83

6.1	DEVELOPMENTS OF RECO HARD MAGNETIC MATERIALS	83
6.2	1:5-TYPE PERMANENT MAGNETS	84
6.3	2:17-TYPE PERMANENT MAGNETS	87

viii ■ Contents

	6.3.1 Phase Constituents	88
	6.3.2 Fabrication Procedures	89
	6.3.3 Latest Developments	91
REFERENCES		95
FURTHER READING		100

CHAPTER 7 ■ **Rare-Earth-Based Hard Magnetic Materials: NdFeB** — 101

7.1	INTRODUCTION TO NdFeB	101
7.2	RESEARCH FOCUSES OF NdFeB	102
	7.2.1 High-Coercivity REFeB with Less Dy/Tb	103
	7.2.2 Low-Cost REFeB with More La/Ce/Y	109
7.3	FABRICATION OF NdFeB SINTERED MAGNETS	113
	7.3.1 Strip Casting	113
	7.3.2 Hydrogen Decrepitation	114
	7.3.3 Jet Milling	114
	7.3.4 Alignment and Compressing	115
	7.3.5 Sintering	116
	7.3.6 Post-Sinter Annealing	118
	7.3.7 Commercial NdFeB Sintered Magnets	120
7.4	FABRICATION OF NdFeB BONDED MAGNETS	120
7.5	SURFACE COATING	121
REFERENCES		122
FURTHER READING		128

CHAPTER 8 ■ **Other Emerging Hard Magnetic Materials** — 129

8.1	NANOCOMPOSITES	129
	8.1.1 Introduction	129
	8.1.2 Timeline of Nanocomposites Development	129
	8.1.3 Synthesis Techniques	131
	8.1.4 Challenges and Perspectives	132

8.2	SmFeN	135
	8.2.1 Crystal Structure	135
	8.2.2 Nitriding Mechanism	136
	8.2.3 Preparation Techniques	138
8.3	Mn-BASED ALLOYS	139
	8.3.1 MnBi Alloy	140
	8.3.2 MnAl and MnAlC Alloys	142
REFERENCES		144
FURTHER READING		150

PART III Soft Magnetic Materials

CHAPTER 9 ▪ Introduction to Soft Magnetic Materials — 153
9.1	APPLICATIONS OF SOFT MAGNETIC MATERIALS	153
9.2	PERFORMANCE REQUIREMENTS FOR SOFT MAGNETIC MATERIALS	154
9.3	DEVELOPMENT OF SOFT MAGNETIC MATERIALS	154
REFERENCES		157
FURTHER READING		158

CHAPTER 10 ▪ Soft Magnetic Alloys — 159
10.1	CRYSTALLINE MAGNETIC ALLOYS	159
	10.1.1 FeSi-Based Magnetic Alloys	159
	10.1.2 FeNi-Based Magnetic Alloys	160
	10.1.3 Future Design of Crystalline Magnetic Alloys	161
10.2	AMORPHOUS MAGNETIC ALLOYS	162
10.3	NANOCRYSTALLINE MAGNETIC ALLOYS	164
	10.3.1 Finemet (FeCuNbSiB)	164
	10.3.2 Nanoperm (FeMBCu, M = Zr, Nb, Hf)	166
	10.3.3 Hitperm (FeCoMBCu, M = Zr, Nb, Hf)	167
	10.3.4 Nanomet (FeSiBPCu)	167
	10.3.5 Future Design of Nanocrystalline Magnetic Alloys	169

REFERENCES	169
FURTHER READING	173

CHAPTER 11 ▪ Soft Magnetic Composites — 175

11.1 POWDER PRODUCTION AND SIZE DISTRIBUTION	176
11.2 INSULATION COATING	177
11.2.1 Organic Coatings	177
11.2.2 Inorganic Coatings	179
11.2.3 Hybrid Organic-Inorganic Coatings	181
11.3 COMPACTION	182
11.4 ANNEALING	183
11.5 CHALLENGES AND PERSPECTIVES	184
REFERENCES	185
FURTHER READING	188

CHAPTER 12 ▪ Soft Magnetic Ferrites — 189

12.1 BASICS OF SOFT MAGNETIC FERRITES	189
12.2 CRYSTAL STRUCTURE OF SOFT FERRITES	190
12.3 POWER LOSS OF SOFT FERRITES	194
12.4 APPLICATIONS OF SOFT FERRITES	196
12.4.1 High Frequency Ferrites	197
12.4.2 High Permeability Ferrites	197
12.4.3 Power Ferrites	198
12.5 MANUFACTURING TECHNOLOGY OF SOFT FERRITES	200
12.6 FUTURE PERSPECTIVES OF SOFT FERRITES	201
REFERENCES	203
FURTHER READING	206

PART IV Other Functional Magnetic Materials

CHAPTER 13 ■ Materials with Magnetic-X Effects — 209

- 13.1 MAGNETO-OPTICAL MATERIALS — 209
  - 13.1.1 Magneto-Optical Effects — 209
  - 13.1.2 Materials Based on Magneto-Optical Effects — 211
    - 13.1.2.1 Magneto-Optical Glass — 211
    - 13.1.2.2 Magneto-Optical Crystals — 212
    - 13.1.2.3 Magneto-Optical Ceramics — 213
  - 13.1.3 Applications — 214
    - 13.1.3.1 Magneto-Optical Recording — 214
    - 13.1.3.2 Magneto-Optical Modulator — 214
    - 13.1.3.3 Magneto-Optical Isolator — 215
    - 13.1.3.4 Magneto-Optical Switcher — 215
- 13.2 MAGNETOSTRICTIVE MATERIALS — 216
  - 13.2.1 Magnetostriction — 216
  - 13.2.2 Materials Based on Magnetostrictive Effects — 217
    - 13.2.2.1 Terfenol-D — 218
    - 13.2.2.2 Galfenol — 219
  - 13.2.3 Applications — 220
- 13.3 MAGNETOCALORIC MATERIALS — 221
  - 13.3.1 Magnetocaloric Effect — 221
  - 13.3.2 Materials Based on Magnetocaloric Effect — 224
    - 13.3.2.1 Gd-Based Alloys — 224
    - 13.3.2.2 Mn-Based Alloys — 225
    - 13.3.2.3 Heusler Alloys — 226
    - 13.3.2.4 LaFeSi Alloys — 226
- REFERENCES — 227
- FURTHER READING — 234

Chapter 14 ▪ Magnetic Materials for Electromagnetic Wave Absorption   235

- 14.1 INTRODUCTION TO ELECTROMAGNETIC WAVE ABSORPTION   235
  - 14.1.1 Impedance Matching   236
  - 14.1.2 Attenuation Capability   237
    - *14.1.2.1 Dielectric Loss*   238
    - *14.1.2.2 Magnetic Loss*   238
  - 14.1.3 Evaluation of the Absorption Performance   239
- 14.2 DEVELOPMENTS OF MAGNETIC ABSORBERS   241
  - 14.2.1 Ferrites for Electromagnetic Wave Absorption   241
  - 14.2.2 Metallic Magnetic Composites for Electromagnetic Wave Absorption   242
- 14.3 FUTURE WORK AND PERSPECTIVES   247
- REFERENCES   247
- FURTHER READING   252

Chapter 15 ▪ Magnetic Materials for Biomedicine, Catalysis and Others   253

- 15.1 MAGNETIC MATERIALS FOR BIOMEDICINE   253
  - 15.1.1 Magnetic Targeting   253
  - 15.1.2 Magnetic Resonance Imaging   254
  - 15.1.3 Magnetic Particle Imaging   256
  - 15.1.4 Magnetic Hyperthermia Therapy   256
- 15.2 MAGNETIC MATERIALS FOR CATALYSIS   259
  - 15.2.1 Magnetic Separation for Catalyst Recycling   259
  - 15.2.2 Direct Involvement of Magnetic Materials in the Catalytic Process   259
  - 15.2.3 Indirect Involvement of Magnetic Materials in the Catalytic Process   260
- 15.3 MAGNETIC MATERIALS FOR OTHER AREAS   264
  - 15.3.1 Micro-Magnetic Robots   264

	15.3.2 Magnetic Fluids and Magnetic Fluidic Platform	266
	15.3.3 Magneto-Electric Vibration Sensor	268
15.4	SUMMARY AND PERSPECTIVES	268
REFERENCES		269
FURTHER READING		273

Author Bio

Chen Wu is an Associate Professor of the School of Materials Science and Engineering at Zhejiang University. She completed a D.Phil. and did post-doctorate research in Materials Science at the University of Oxford. Her research interest lies in the design, fabrication and manipulation of magnetic composites and their wide frequency applications.

Jiaying Jin is an Associate Professor at the School of Materials Science and Engineering at Zhejiang University. Her main research interests are magnetism and permanent magnetic materials.

I

Fundamental Magnetism

CHAPTER 1

Introduction to Magnetics

1.1 INTRODUCTION TO MAGNETISM AND MAGNETIC MATERIALS

Magnetism is delicate as it cannot be seen or touched. It is, however, ubiquitous since the planet we live is a huge magnet due to the slow cooling of the liquid iron at the core of the earth (Davies et al., 2015). Despite of the small geomagnetic field (around 1/100 of that generated by a fridge magnet), it is critical to life on earth. Firstly, the geomagnetism protects us from cosmic radiation since it forms a magnetosphere consisting of charged particles, which deflects the solar wind from the earth. Secondly, geomagnetism has long been used for guidance of direction. Migrants such as turtles, whales and birds travel thousands of kilometers every year without getting lost thanks to their bio-sensitivity to the direction and strength of the geomagnetic field. Back in the period 206 BC–AD 220, the south pointer was invented based on the lodestone (Fe_3O_4) (Coey, 2010). This was later modified by Kuo Shen into the navigational compass in the 1100s, which later assisted in the discovery of Africa by He Zheng, and other great discoveries of new lands by Christopher Columbus, Vasco da Gama and Ferdinand Magellan after the 1400s. Thirdly, the geomagnetic field also aids mining for iron, nickel, chromium, etc., which also generate fields to alter the distribution of the geomagnetic field. If "magnetic minerals" are deformed under stress, their magnetism may change, based on which forecast of major earthquakes become possible.

Not only is magnetism closely related to our daily life, but magnetic materials have also become increasingly important since their first application in the compass. Magnetic materials have now ranked the second only to semiconductors as the world's largest industry, accounting for a global market of more than US $30 billion annually (Coey, 2010). The family of magnetic materials can be categorized into hard magnets, soft magnets and other functional magnetic materials, which has been supporting the development of electronic devices, communication technologies, transportation and even biomedicine. For instance, hard magnets providing large flux and stable field are used in electric motors and nuclear magnetic resonance imaging, whereas soft magnets with rapid response to external field are critical for power conversion as in generators, transformers and inductors. During the last century, tremendous progress has been achieved for hard and soft magnets. In particular, energy products of hard magnetic materials have doubled every 14 years with the development of new hard magnetic materials, while the power loss of the soft magnetic materials has halved every 18 years due to the advances in the fabrication technologies (Coey, 2010). Other functional magnetic materials have also been developing diversely and interdisciplinary, with the emergency of various smart materials with magnetic-optical, electronic, mechanical or caloric effects. The applications of the magnetic materials have also been expanded to multiple fields such as biomedicine and catalysis. To date, the ever-rising demands from both conventional and emerging applications continuously push the advances in magnetic materials. To fulfill such demands, in-depth understanding of the principles of magnetism, as well as correlations between the composition, microstructure and magnetic performance is essential.

Figure 1.1 demonstrates the transition from classical magnetism to micromagnetism and quantum magnetism due to the development of in-depth understanding at increasingly delicate scales. The classical magnetism describes magnetic performance through characterization of the bulk magnets at the macroscale which then evolves into the microscale with the investigation of magnetic domains. Further development lies in the micromagnetism, covering the magnetization process of ferromagnets at the scale between 1 nm–1 μm, including the sub micro- (100 nm–1 μm) and nano- (1 nm–100 nm) scales. At the submicroscale, magnetically ordered area forms magnetic domains which may be extendable

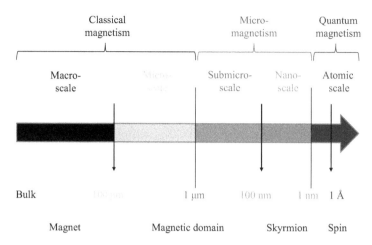

FIGURE 1.1 Transition from classical magnetism to micromagnetism and quantum magnetism.

into the microscale. Whereas at the nanoscale, the skyrmions, topologically nontrivial magnetic swirls with bended magnetic moments have become an important research direction for data storage (Tang et al., 2021). Quantum magnetism at the atomic scale unveils the origin of magnetic moments, involving the spin and orbital movements of the electrons. The evolution from classical to quantum magnetism relies on the advances in the measurement techniques which enable us to "see" the nanoscale structures of the materials and to "sense" the magnetic resonance generated by electron movements. With the assistance of such powerful tools which will be described in detail in Chapter 3, we are able to tailor the macro-properties by finely tuned atomic arrangement and electron spins.

1.2 VIEWS ON MAGNETISM AND TWO UNIT SYSTEMS

In this book, we shall discuss fundamental magnetisms, magnetic measurement techniques as well as the latest development of various magnetic materials including hard magnets, soft magnets and other functional magnetic materials. To bridge these aspects, important magnetic parameters to evaluate the performance will be described. Unlike most of the physical parameters described in international system (SI), two unit systems are prevailing in magnetism depending on the understanding perspectives. Early application of natural magnets leads to an engineer's perspective, where magnetism is generated by magnetic poles, giving rise to the centimeter-gram–second (CGS) system.

The other perspective originates from a milestone in magnetism which is the discovery of electromagnetic conversion. In 1820, Hans-Christian Oersted demonstrated deflection of a compass needle by a wire with running current (Spaldin, 2010), while in 1821, Michael Faraday discovered electromagnetic induction (Coey, 2010). Later on, James Clerk Maxwell put forward the famous Maxwell's Equations as follows which unify the theory of electricity and magnetism. The simplified physical meanings corresponding to the formulas as in Equations (1.1)–(1.4) include (i) the magnetic field is sourceless and line of magnetic force forms a close loop; (ii) charge is the source of electric field; (iii) current and changing electric field generate magnetic field; and (iv) changing magnetic field gives rise to electric field.

$$\nabla \cdot B = 0 \qquad (1.1)$$

$$\varepsilon_0 \nabla \cdot E = \rho \qquad (1.2)$$

$$(1/\mu_0)\nabla \times B = j + \varepsilon_0 \partial E / \partial t \qquad (1.3)$$

$$\nabla \times E = -\partial B / \partial t \qquad (1.4)$$

TABLE 1.1 Unit Conversions for the Frequently Used Magnetic Parameters

Name		CGS Symbol and Unit		SI Symbol and Unit	Conversion
Magnetic flux	Φ	Maxwell (Mx)	Φ	Weber (Wb)	1 Mx = 10^{-8} Wb
Magnetic flux density	B	Gaussian (Gs)	B	Tesla (T)	1 Gs = 10^{-4} T
Magnetic field strength	H	Oersted (Oe)	H	Ampere/Meter (A/m)	1 Oe = $10^3/4\pi$ A/m
Magnetization	M	Gaussian (Gs)	M	Ampere/Meter (A/m)	1 Gs = 10^3 A/m
Magnetic polarization	$4\pi M$	Gaussian (Gs)	J	Tesla (T)	1 Gs = 10^{-4} T
Magnetic energy product	BH	Gaussian·Oersted (GOe)	BH	Joules/Meter^3 (J/m^3)	1 MGOe = $10^2/4\pi$ kJ/m^3

The discovery of the connections between electricity and magnetism gives rise to the physicists' view, where magnetism originates from circulating currents. This gives rise to the SI unit. The two unit systems for magnetism have both been used by engineering workers and scientific researchers worldwide. Specifically, for the evaluation of the magnetic performance during production or application as in the "engineering occasions", the CGS unit system is frequently used, while for fundamental understanding of magnetisms as in the "scientific world", the SI unit system is likely to be used. Since this book intends to bridge the basics of magnetism and the design/applications of magnetic materials, we may use both unit systems according to whether it is to describe the concept based on interactions between "magnetic poles" or "circular currents". Conversion between the two unit systems are summarized in Table 1.1 for future reference.

REFERENCES

Coey, J. M. D. 2010. *Magnetism and magnetic materials*. New York: Cambridge University Press.

Davies, C., M. Pozzo, D. Gubbins, and D. Alfe. 2015. Constraints from material properties on the dynamics and evolution of Earth's core. *Nature Geoscience* 8:678–85.

Spaldin, N. A. 2010. *Magnetic materials: fundamentals and applications*. New York: Cambridge University Press.

Tang, J., Y. Wu, W. Wang, L. Kong, B. Lv, W. Wei, J. Zang, M. Tian, and H. Du. 2021. Magnetic skyrmion bundles and their current-driven dynamics. *Nature Nanotechnology* 16:1086–91.

FURTHER READING

Gutfleisch, O., M. A. Willard, E. Brück, C. H. Chen, S. G. Sankar, and J. P. Liu. 2011. Magnetic materials and devices for the 21st century: stronger, lighter, and more energy efficient. *Advanced Materials* 23:821–42.

CHAPTER 2

Origin and Categories of Magnetisms

2.1 ATOMIC ORIGIN OF MAGNETISMS

Magnetism is in fact a fundamental feature for all kinds of materials which are constructed by atoms containing nucleus and electrons. Based on the classical theory of electromagnetic conversion, the movement of charged particles induces magnetism. Magnetism may then stem from the movement of both atomic nucleus and electrons. Given the fact that the magnetic moment of nucleus is very small and negligible, the magnetism mainly originates from the electron orbital moment μ_l and electron spin moment μ_s as shown in Figure 2.1.

2.1.1 Arrangement of Electrons

The status of each electron is governed by four quantum numbers including the principal quantum number (n = 1, 2, 3, 4… represented by K, L, M, N…) to determine the energy level of the electron, the angular momentum quantum number (l = 0, 1, 2, 3… n–1 represented by s, p, d, f…) to determine the magnitude of the orbital angular momentum, the magnetic quantum number (m_l = 0, ±1… ±l) to determine the orientation of the orbital angular momentum with respect to the magnetic field, and the magnetic spin quantum number (m_s = ±1/2) for quantized spin angular momentum. For a many-electron atom, the arrangement of the electrons follows the Pauling exclusion principle, stating that no two electrons

DOI: 10.1201/9781003216346-3

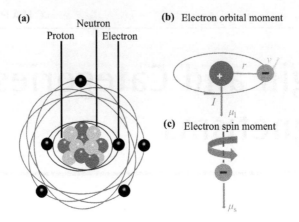

FIGURE 2.1 (a) Configuration of the atom, (b) electron orbital moment μ_l and (c) electron spin moment μ_s.

have exactly the same set of quantum numbers (Pauli, 1946). The electrons usually fill the shells from low to high energy levels, following the sequence of 1s, 2s, 2p, 3s, 3p... Due to the energy-level-interlaced problem, however, from the 4s subshell, the s subshells are preferentially filled prior to the d subshells. Such issue also applies to the filling of 5p and 6s subshells prior to the 4f subshell. The actual arrangement of the electrons can be described as $1s^2\ 2s^2\ 2p^6\ 3s^2\ 3p^6\ 4s^2\ 3d^{10}\ 4p^6\ 5s^2\ 4d^{10}\ 5p^6\ 6s^2\ 4f^{14}\ 5d^{10}\ 6p^6\ 7s^2\ 5f^{14}\ 6d^{10}\ 7p^6$, which is well illustrated in the sequence following the dashed arrows in Figure 2.2.

Generally, the net magnetic moment originates from unpaired electrons, so the "magnetic" atoms usually refer to the elements with unfilled 3d and 4f subshells. Table 2.1 summarizes the electron arrangements for the 3d and 4f atoms. It is worth noting that the electron configuration of the outer shells is $3d^5\ 4s^1$ instead of $3d^6\ 4s^2$ for the Cr, and $3d^{10}\ 4s^1$ rather than $3d^9\ 4s^2$ for the Cu as highlighted in bold in the table. This can be explained by the supplementary Hund's rule that lower energy can be obtained when the shell is empty, half-filled and completely filled (Masatoshi et al., 1998). Such rule also applies to the 4f atoms including La and Gd with $5d^1\ 6s^2$ and $4f^7\ 5d^1\ 6s^2$ configuration, while the electron arrangement of the Ce atom is an exception.

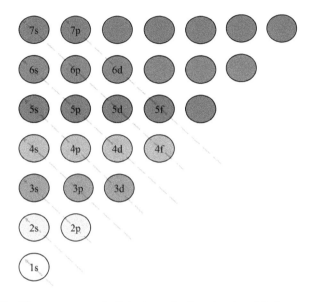

FIGURE 2.2 The arrangement of electrons in the atomic orbitals.

TABLE 2.1 Arrangement of Electrons for 3d and 4f Atoms

3d atoms

20 Ca calcium: [Ar] 4s2
21 Sc Scandium: [Ar] 3d1 4s2
22 Ti titanium: [Ar] 3d2 4s2
23 V vanvadium: [Ar] 3d3 4s2
24 Cr chromium: [Ar] 3d5 4s1
25 Mn manganse: [Ar] 3d5 4s2

26 Fe iron: [Ar] 3d6 4s2
27 Co cobalt: [Ar] 3d7 4s2
28 Ni nickel: [Ar] 3d8 4s2
29 Cu copper: [Ar] 3d10 4s1
30 Zn zinc: [Ar] 3d10 4s2

4f atoms

57 La lanthanum: [Xe] 5d1 6s2
58 Ce cerium: [Xe] 4f1 5d1 6s2
59 Pr praseodymium: [Xe] 4f3 6s2
60 Nd neodymiun: [Xe] 4f4 6s2
61 Pm promethium: [Xe] 4f5 6s2
62 Sm samarium: [Xe] 4f6 6s2
63 Eu europium: [Xe] 4f7 6s2
64 Gd gadolinum: [Xe] 4f7 5d1 6s2
65 Tb terbium: [Xe] 4f9 6s2

66 Dy dysprosium: [Xe] 4f10 6s2
67 Ho holmiun: [Xe] 4f11 6s2
68 Er erbium: [Xe] 4f12 6s2
69 Tm thulium: [Xe] 4f13 6s2
70 Yb ytterbium: [Xe] 4f14 6s2
71 Lu lutetium: [Xe] 4f14 5d1 6s2
72 Hf hafnium: [Xe] 4f14 5d2 6s2
72 Ta tantalum: [Xe] 4f14 5d3 6s2

2.1.2 Electron Orbital and Spin Moment

The orbital angular moment of the electrons μ_l reflects Ampère's idea about circulating currents, expressed as

$$\mu_l = -\frac{e}{2m_e} p_l \tag{2.1}$$

where e and m_e are the charge and mass of the electron, and p_l is the orbital angular momentum. The orbital angular moment of the electron is quantized and its magnitude $|L|$ depends on the angular momentum quantum number l with the correlation of

$$|L| = \sqrt{l(l+1)}\,\hbar \tag{2.2}$$

where $\hbar$ is the reduced Planck constant. Figure 2.3 shows the quantized direction of the orbital angular momentum with respect to the magnetic field H. For instance, for the p orbital with $l = 1$, the m_l can be −1, 0 and +1, while for the d orbital with $l = 2$, the m_l values at −2, −1, 0, +1, +2, etc.

Similarly, the spin moment μ_s is contributed by the intrinsic electron angular momentum p_s given by

$$\mu_s = -\frac{e}{m_e} p_s \tag{2.3}$$

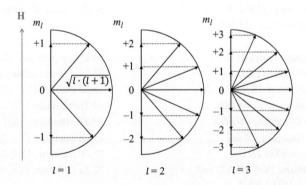

FIGURE 2.3 The direction of the orbital angular momentum with respect to the magnetic field H.

The spin quantum number s of the electron possesses the value of $\pm 1/2$, and the magnitude of the spin angular momentum $|S|$ follows

$$|S| = \sqrt{s(s+1)}\,\hbar \tag{2.4}$$

2.1.3 Total Magnetic Moment of the Atom

The Hund's rules determine the lowest-energy configuration for the electrons (Hund, 1927), stating that (i) the total spin S is maximized in the electrons; (ii) for a given spin arrangement, largest total atomic orbital angular momentum L is preferential; and (iii) atoms with less than half-full shells exhibit the smallest energy with the lowest value of J (i.e. $J = |L - S|$), while atoms with more than half-full shells, the opposite rule stands with the lowest energy achieved at the highest J ($J = |L+S|$). For the case of exactly half-filled shell, $L = 0$ and $J = S$.

The total atomic moment incorporates both orbital and spin angular momentum and can be calculated based on their interaction, i.e. the spin-orbit coupling. The total atomic moment is proportional to the projection of J, and is expressed as

$$\mu_J = -g\frac{e}{2m_e}J \tag{2.5}$$

where g is the spectroscopic splitting factor, and the magnitude of J is described by

$$|J| = \sqrt{J(J+1)}\,\hbar \tag{2.6}$$

Combining Equations (2.5) and (2.6), we have

$$|\mu_J| = g\frac{e}{2m_e}\sqrt{J(J+1)}\,\hbar = g\mu_B\sqrt{J(J+1)} \tag{2.7}$$

with $\mu_B = \dfrac{e}{2m_e}$ as the Bohr magneton.

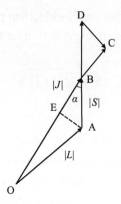

FIGURE 2.4 Scheme of the *L–S* coupling.

Figure 2.4 shows the geometry of the *L–S* coupling (Krishnan, 2016). For the line OBC, we have

$$g|J| = |J| + |S|\cos\alpha \qquad (2.8)$$

For the triangle AOE, we get

$$|L|^2 = (|J| - |S|\cos\alpha)^2 + (|S|\sin\alpha)^2 \qquad (2.9)$$

$$|L|^2 = |J|^2 + |S|^2 \cos^2\alpha - 2|J||S|\cos\alpha + |S|^2 \sin^2\alpha \qquad (2.10)$$

$$|L|^2 = |J|^2 + |S|^2 - 2|J|(g|J| - |J|) \qquad (2.11)$$

$$g = 1 + \frac{|J|^2 + |S|^2 - |L|^2}{2|J|^2} \qquad (2.12)$$

Based on Equations (2.2), (2.4) and (2.6), we have

$$g = 1 + \frac{J(J+1) + S(S+1) - L(L+1)}{2J(J+1)} \qquad (2.13)$$

The magnetic moment of 3d and 4f ions can be calculated by the formulas above. An example is given by estimating the moment of the Fe^{2+}.

TABLE 2.2 Arrangement of 3d Electrons for the Fe^{2+} Ion

m_l	2		1		0		−1		−2	
m_s	1/2	−1/2	1/2	−1/2	1/2	−1/2	1/2	−1/2	1/2	−1/2
Electron number	1	1	1		1		1		1	

The Fe possesses an atomic number of 26, and the arrangement of electrons is 1s² 2s² 2p⁶ 3s² 3p⁶ 4s² 3d⁶. For the Fe^{2+} ion, 2 electrons at the 4s subshell are lost, with 6 electrons remaining in the 3d orbital. To satisfy the Hund's rules described above, detailed arrangement of the 3d electrons is illustrated in Table 2.2.

We have the total spin quantum number of the Fe^{2+} ion as $S = 5 \times 1/2 - 1/2 = 2$ and total orbital quantum number $L = 2 + 1 + 0 - 2 - 1 + 2 = 2$. Since the 3d shell is more than half-full, $J = |L + S| = 4$. According to Equation (2.13), g of the Fe^{2+} can be calculated as 1.5, giving rise to the μ_J of 6.70 μ_B based on Equation (2.7). The experimentally measured μ_J of the Fe^{2+} is, however, 5.36 μ_B (Krishnan, 2016). This can be explained by quenching of the orbital angular momentum where electric field originated from the surrounding ions results in strong coupling of the orbitals to the lattice. In fact, for 3d transition metal ions, since the 3d electrons are the outermost electrons that participate in bonding and are strongly affected by the crystal lattice, the orbital angular momentum tends to be quenched or partially quenched. As a result, the actual magnetic moment of many 3d metals is different from that calculated based on the L–S coupling model.

For the 4f metal ions, since the 4f electrons are well under the 5s and 5p orbitals that are involved in bonding, the L–S coupling surpasses the crystal field effect. Consequently, little discrepancy is resulted between the calculated and measured magnetic moment. An example is given by calculation of the moment for the Nd^{3+} ion. The Nd has an atomic number of 60. Considering the loss of 3 electrons to form the Nd^{3+}, the electron arrangement becomes 1s² 2s² 2p⁶ 3s² 3p⁶ 4s² 3d¹⁰ 4p⁶ 5s² 4d¹⁰ 5p⁶ 4f³ with 3 electrons in the 4f orbital as shown in Table 2.3.

TABLE 2.3 Arrangement of 4f Electrons for the Nd^{3+} Ion

m_l	3		2		1		0		−1		−2		−3	
m_s	1/2	−1/2	1/2	−1/2	1/2	−1/2	1/2	−1/2	1/2	−1/2	1/2	−1/2	1/2	−1/2
Electron number	1		1		1									

Similarly, we have $S = 3 \times 1/2 = 3/2$ and $L = 3 + 2 + 1 = 6$. Since the 4f shell contains electrons less than half-full, $J = |L - S| = 9/2$. Again, according to Equation (2.13), g of the Nd^{3+} can be calculated as 0.73. Based on Equation (2.7), the μ_J of the Nd^{3+} ion is estimated as 3.62 μ_B.

2.2 FIVE TYPES OF MAGNETISMS

As discussed above, each electron in an atom possesses its own spin axes and moving orbitals, giving rise to magnetic moment as a vector parallel to the spin axis and normal to the orbital plane. The vector sum of all the electronic moments results in the overall magnetic moment. If all the moments of the electrons cancel out with each other, diamagnetism arises with the atom exhibiting zero net magnetic moment. If the electronic moments are only partially canceled out, a net magnetic moment is resulted, which may produce several possible magnetisms including paramagnetism, antiferromagnetism, ferromagnetism and ferrimagnetism depending on the interactions of the magnetic moments as shown in Figure 2.5. Details of these types of magnetisms will be introduced in the following sections.

2.2.1 Diamagnetism

Diamagnetism is a common phenomenon for all substances, which can be understood based on electromagnetic induction. An applied field changes the magnetic flux of the electron orbit, which induces alteration

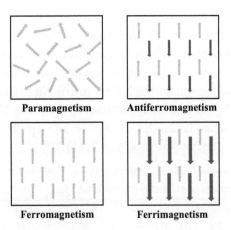

FIGURE 2.5 Ordering of magnetic moment for paramagnetism, antiferromagnetism, ferromagnetism and ferrimagnetism.

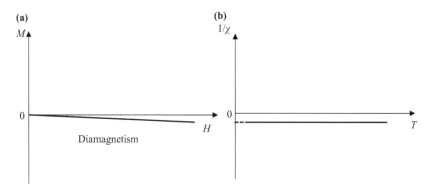

FIGURE 2.6 The (a) M–H and (b) $1/\chi$–T of a typical diamagnetic material.

of the orbital motion to repelled such change. This results in a negative linear correlation between the magnetization M and the external field H (Figure 2.6a), as well as a small and negative susceptibility $\chi = M/H$, which is independent of temperature (Figure 2.6b). The diamagnetic behavior is a weak phenomenon and substances without net magnetic moment due to completely filled electronic shells are usually considered as diamagnetic materials, such as noble gases and ionic solids. For ferrimagnetic and ferromagnetic materials, the diamagnetism is usually neglectable due to the much stronger magnetic moment interactions.

One well-known material that exhibits diamagnetic behavior is the superconductor for which the electrical resistivity becomes zero when cooled below a critical temperature T_C (Onnes, 1991). The superconductor is in fact a "perfect" diamagnetic material below the T_C. It expels all magnetic flux from inside with $B = \mu_0(H + M) = 0$, giving rise to $M = -H$. Since $\chi = M/H = -1$ and $\mu = 1 + \chi = 0$, indicating that the material is not permeable for the applied magnetic field. Such behavior is resulted from the opposing of applied field by the macroscopic currents circulating in the superconductor, rather than the changing orbital motion of the moving electrons for conventional diamagnetic materials.

2.2.2 Paramagnetism

Paramagnetism occurs for materials with weak-coupled and random-aligned magnetic moments due to thermal motions. Under an external magnetic field, only a small proportion of the moments align along the field direction, resulting in linear correlation between the M and H with

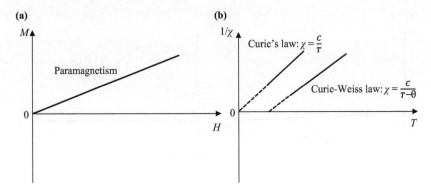

FIGURE 2.7 The (a) M–H and (b) $1/\chi$–T curves of typical paramagnetic materials with and without consideration of the molecular field as in the Curie-Weiss law and the Curie's law.

a small and positive slope coefficient (Figure 2.7a). The classical Langevin theory of paramagnetism is given by (Cabrera, 1927)

$$M = NmL(\alpha) \tag{2.14}$$

where N is the number of atoms or particles in the system, m is the magnitude of the magnetic moment, $\alpha = \dfrac{mH}{k_B T}$ and the Langevin function gives $L(\alpha) = \coth \alpha - 1/\alpha$. At low field or high temperature, $\alpha \ll 1$ and $\coth \alpha = 1/\alpha + \alpha/3 - \alpha^3/45 + \ldots$
Then we get

$$M = \frac{Nm\alpha}{3} = \frac{Nm^2 H}{3k_B T} \tag{2.15}$$

This gives the following relationship between the susceptibility and temperature as in the Curie's law (Figure 2.7b), assuming no interaction between the magnetic moments.

$$\chi = \frac{M}{H} = \frac{Nm^2}{3k_B T} = \frac{C}{T} \tag{2.16}$$

According to the Weiss theory, interaction exists between the localized moments and generates a molecular field. The intensity of the Weiss field H_w is given by (Weiss, 1907)

$$H_W = qM \tag{2.17}$$

where q is the molecular field constant. The total field then follows

$$H_{tot} = H + H_W = H + qM \tag{2.18}$$

Replacing it to Equation (2.15), we have

$$M = \frac{Nm^2(H+qM)}{3k_BT} \tag{2.19}$$

Rearranging the terms, we get

$$\left[\left(\frac{3k_BT}{\mu_0 Nm^2}\right) - q\right]M = H \tag{2.20}$$

$$\chi = \frac{M}{H} = \frac{\mu_0 Nm^2}{3k_BT - \mu_0 Nm^2 q} = \frac{C}{T-qC} = \frac{C}{T-\theta} \tag{2.21}$$

Changes of the $1/\chi$ in correspondence of T for paramagnetic materials with the consideration of the molecular field as in the Curie-Weiss law is also plotted in Figure 2.7b.

Typical paramagnetic materials include metals such as aluminum and platinum, gases like oxygen and nitric oxide, as well as some salts of transition elements or rare earth elements. It is worth noting that for ferromagnetic and ferrimagnetic materials above their Curie temperatures, the thermal energy k_BT exceeds the alignment achieved by the externally applied magnetic field, resulting in transition into paramagnetism. Also, for ferromagnetic and ferrimagnetic substances with reduced size, decreased anisotropy energy KV is resulted where K and V are the anisotropy constant and the particle volume, respectively. If the KV with reduced V decreases to approach the thermal energy k_BT, the thermal motion of particles may surpass the anisotropy and make the alignment of the particles randomly. Such phenomenon is called superparamagnetism where small particles exhibit similar but larger magnetization compared to that of the paramagnetic materials.

2.2.3 Antiferromagnetism

Antiparallel-aligned adjacent magnetic moments with equal magnitude give rise to antiferromagnetism. Similar to that of the paramagnetic materials, antiferromagnetic materials exhibit positive linear M–H correlation as shown in Figure 2.8a. The first evidence for antiferromagnetism was generated by neutron diffraction spectrum of the MnO back to 1949, where two sets of antiparallel spins were observed for the Mn^{2+} (Shull and Smart, 1949). Actually, most antiferromagnetic materials are ionic compounds such as oxides, sulfides and chlorides where super-exchange occurs involving O, S and Cl mediated interactions between the metal ions with upward and downward spins (Coey, 2010).

Although the antiferromagnetic substances exhibit a small positive susceptibility, it varies with temperature as shown in Figure 2.8b. In order to understand this, the Weiss theory is also considered here with two internal molecular fields given by

$$H_{wA} = -q_{AB} M_B \quad (2.22)$$

$$H_{wB} = -q_{AB} M_A \quad (2.23)$$

where $H_{wA(B)}$ is the Weiss field and $M_{A(B)}$ is the magnetization for each of the two sub-lattices including A and B. When it is above the Néel temperature T_N, the Curie's law gives

$$\chi = \frac{C}{T} = \frac{M}{H} \quad (2.24)$$

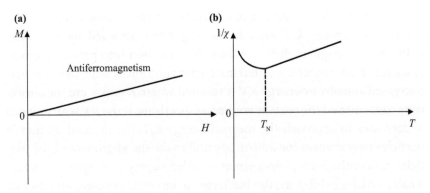

FIGURE 2.8 The (a) M–H and (b) $1/\chi$–T of typical antiferromagnetism.

where H consists of both the applied and the Weiss field. This gives rise to

$$M_A T = C'(H + H_{wA}) = C'(H - qM_B) \tag{2.25}$$

$$M_B T = C'(H + H_{wB}) = C'(H - qM_A) \tag{2.26}$$

Adding Equations (2.25) and (2.26), we have

$$(M_A + M_B)T = 2C'H - qC'(M_A + M_B) \tag{2.27}$$

By definition, $M_A + M_B = M$ as the total magnetization for $T > T_N$. Therefore, we have

$$\chi = \frac{C}{T} = \frac{M}{H} = \frac{2C'}{T + qC'} = \frac{C}{T + \theta} \tag{2.28}$$

Below the T_N, if the applied field is parallel to magnetization, the susceptibility of the antiferromagnetic materials is given by (Spaldin, 2010)

$$\chi_\parallel = \frac{2Nm^2 B'(J,a)}{2k_B T + Nm^2 \gamma B'(J,a)} \tag{2.29}$$

where N and $B'(J,a)$ are the atomic number in the unit volume and the Brillouin function derivative, respectively. With perpendicularly applied field, the susceptibility of the antiferromagnetic materials is given by

$$\chi_\perp = \frac{M}{H} = \frac{1}{\gamma} \tag{2.30}$$

Consequently, for temperatures below T_N, the total susceptibility of the antiferromagnetic materials is the vector sum of the $\chi_\parallel$ and $\chi_\perp$ as shown in Figure 2.8b.

2.2.4 Ferromagnetism

In ferromagnetic materials, individual atom has a net magnetic moment with strong interactions with each other. The response of such materials

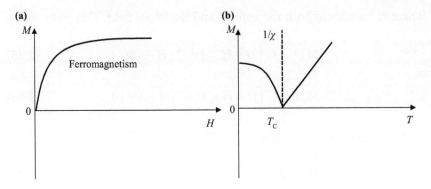

FIGURE 2.9 The (a) M–H curve and (b) M and $1/\chi$ as a function of T for typical ferromagnetism.

to an external field is non-linear (Figure 2.9a) with hysteresis depending on the composition, microstructure and other extrinsic factors. The relationship between spontaneous magnetization and the temperature of the ferromagnetism are shown in the left region of Figure 2.9b. According to the Weiss theory, the total field is given by

$$H_{tot} = H + H_w = H + qM \quad (2.31)$$

For relatively large M of the ferromagnetic material, higher-order terms need to be considered, giving rise to the total field as follows (Krishnan, 2016)

$$H_{tot} = H + H_w = H + \left(q - bM^2\right)M \quad (2.32)$$

Following the Curie's law, we have

$$M\left(1 - \frac{qC}{T} + \frac{CbM^2}{T}\right) = \frac{CH}{T} \quad (2.33)$$

Assuming that $T_C = qC$, for $T < T_C$ and $H = 0$, we get

$$M\left(1 - \frac{qC}{T} + \frac{CbM^2}{T}\right) = 0 \quad (2.34)$$

So

$$M^2 = \frac{qC-T}{Cb} \tag{2.35}$$

For $T > T_C$, ferromagnetism tends to transform into paramagnetism and similar relationship between the $1/\chi$ and T is obtained as shown in the right region in Figure 2.9b.

Ferromagnetic materials are suitable for wide applications. For instance, hard magnetic materials are frequently incorporated in electric machines to provide stable magnetic flux, while soft magnets with rapid response to applied field are crucial for power generation and conversion. In addition, ferromagnetism may also affect mechanical, thermal and optical properties of solids as in the magnetostriction, magnetocaloric and magneto-optic effect for applications in various sensors.

2.2.5 Ferrimagnetism

Ferrimagnetic materials contain antiparallel-aligned magnetic moments but with unequal magnitudes. The simplest model to describe the features of ferrimagnetism may involve A–A, B–B and A–B ion pair interactions, among which the A–B interactions induce antiparallel alignment, and the A–A and B–B pairs exhibit ferromagnetic interactions with varied moments.

Different from antiferromagnetic materials, ferrimagnetic materials exhibit a net magnetic moment below the Curie temperature, above which they show paramagnetism. The response of ferrimagnetic materials to an external field is similar to that of the ferromagnetic materials (Figure 2.10a). According to the Weiss theory, the inner field of ferrimagnetic material is given by (Krishnan, 2016)

$$H_{wA} = q_{AA}M_A - q_{AB}M_B \tag{2.36}$$

$$H_{wB} = q_{BB}M_B - q_{AB}M_A \tag{2.37}$$

When the temperature is below the T_C, a net magnetization arises as

$$M = |M_A| - |M_B| \tag{2.38}$$

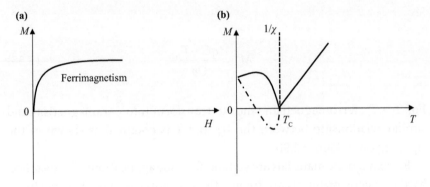

FIGURE 2.10 The (a) M–H curve and (b) M and $1/\chi$ as a function of T for typical ferrimagnetism.

The Brillouin function magnetization gives

$$M_A = Nm_A B_j\left(J, \frac{\mu_0 m H_{wA}}{k_B T}\right) \tag{2.39}$$

$$M_B = Nm_B B_j\left(J, \frac{\mu_0 m H_{wB}}{k_B T}\right) \tag{2.40}$$

where m_A and m_B are the magnetic moments of the A and B ions, respectively. Replacing H_{wA} and H_{wB} from Equations (2.36) and (2.37), we have

$$M_A = Nm_A B_j\left(J, \frac{\mu_0 m[q_{AA} M_A - q_{AB} M_B]}{k_B T}\right) \tag{2.41}$$

$$M_B = Nm_B B_j\left(J, \frac{\mu_0 m[q_{BB} M_B - q_{AB} M_A]}{k_B T}\right) \tag{2.42}$$

The magnetization of both sublattice A and B reduces with raising temperature. If the magnetization of sublattice A decreases less rapidly than that of sublattice B, the net magnetization increases with raised temperature after reaching a maximum where the net magnetization decreases and falls to zero at T_C, as shown by the solid line in the left region in Figure 2.10b. The dashed line illustrates a different case where the net magnetization disappears before the T_C, and the magnetization takes the opposite direction afterward.

When $T > T_C$, according to the Weiss theory, we get (Krishnan, 2016)

$$\frac{M_A}{H_{tot}^A} = \lambda \frac{C}{T} \tag{2.43}$$

$$\frac{M_B}{H_{tot}^B} = \mu \frac{C}{T} \tag{2.44}$$

where λ and μ are the fractions of the two lattices. Rearranging terms, we have

$$M_A T = \lambda C H_{tot}^A = \lambda C(H + H_{wA}) = \lambda C(H + q_{AA} M_B - q_{AB} M_B) \tag{2.45}$$

$$M_B T = \mu C H_{tot}^B = \mu C(H + H_{wB}) = \mu C(H + q_{BB} M_B - q_{AB} M_A) \tag{2.46}$$

Let $\alpha = \frac{q_{AA}}{q_{AB}}$ and $\beta = \frac{q_{BB}}{q_{AB}}$, the susceptibility of the combined two lattices is given by

$$\frac{1}{\chi} = \frac{M}{H} = \frac{T}{c} + \frac{1}{\chi_0} - \frac{\zeta}{T - \theta} \tag{2.47}$$

where

$$\frac{1}{\chi_0} = q_{AB}(2\lambda\mu - \lambda^2\alpha - \mu^2\beta) \tag{2.48}$$

$$\theta = q_{AB}\lambda C(2 + \alpha + \beta) \tag{2.49}$$

$$\zeta = q_{AB}^2 \lambda\mu C\{\lambda(1+\alpha) - \mu(1+\beta)\}^2 \tag{2.50}$$

Ferrimagnetism is normally observed in materials containing two or more lattice positions with different magnetic moments such as spinel, garnet and perovskite compounds. Soft ferrites as an important category of ferrimagnetic materials have been widely used for high-frequency applications due to their combined net magnetization and large electrical resistivity for reduced loss. Further information on the design and fabrication of soft ferrites, refer to Chapter 12.

In summary, magnetisms mainly origin from orbital and spin motions of the electrons. Even for atoms with filled electron shells (i.e. spin up and spin down cancel out each other), the orbital motions tend to repel changes of the external field, giving rise to diamagnetism. Although it applies to all types of atoms, the diamagnetism without any net magnetic moment, is the weakest among the five magnetisms. For paramagnetism, the magnetic moments arrange themselves randomly which deflect to the direction of the applied field, exhibiting a linear M–H relationship. Similar M–H behavior can be obtained for antiferromagnetic materials involving two sets of antiparallel-aligned equal moments. For both randomly and antiparallel arranged moments, they cancel out each other without applied external field. Consequently, paramagnetism and antiferromagnetism are relatively weak as well. The so-called magnetic materials generally refer to ferromagnetic and ferrimagnetic materials. The former contains parallel-aligned moments while the later involves antiparallel-aligned and unequal moments. Since the moments in ferrimagnetism only partially cancel out each other, the uncancelled moments behave similar to those in ferromagnetic materials. The ferromagnetic and ferrimagnetic materials possess relatively strong magnetism and are the most widely used in real life, which will be further discussed in the following chapters.

REFERENCES

Cabrera, B. 1927. La théorie du paramagnétisme. *Journal de Physique et le Radium* 8:257–75.

Coey, J. M. D. 2010. *Magnetism and magnetic materials*. New York: Cambridge University Press.

Hund, F. 1927. *Linienspektren und periodisches system der elemente*. Vienna: Springer.

Krishnan, K. M. 2016. *Fundamentals and applications of magnetic materials*. New York: Oxford University Press.

Masatoshi, I., F. Atsushi, and T. Yoshinori. 1998. Metal-insulator transitions. *Reviews of Modern Physics* 70:1039–263.

Onnes, H. K. 1991. *Further experiments with liquid helium*. Dordrecht: Springer.

Pauli, W. 1946. Remarks on the history of the exclusion principle. *Science* 103:213–15.

Shull, C. G., and J. S. Smart. 1949. Detection of antiferromagnetism by neutron diffraction. *Physical Review* 76:1256–57.

Spaldin, N. A. 2010. *Magnetic materials: fundamentals and applications*. New York: Cambridge University Press.

Weiss, P. 1907. L'hypothèse du champ moléculaire et la propriété ferromagnétique. *Journal de Physique* 6:661–90.

FURTHER READING

Buschow, K. H. J., and F. R. De Boer. 2003. *Physics of magnetism and magnetic materials*. New York: Kluwer Academic Publishers.

CHAPTER 3

Important Parameters and Magnetic Measurements

3.1 BASICS OF MAGNETIZATION AND MAGNETIC PARAMETERS

Ferromagnetic and ferrimagnetic materials are the mostly used magnetic materials exhibiting remarkable responsive characteristics to an applied magnetic field. Their characteristics are frequently described by the magnetization curve and hysteresis loop. Magnetization curves represent the correlations between the magnetic induction B (or magnetization M) and the applied field H. In fact, the B integrates the field strength and the intrinsic magnetization of the material described by

$$B = \mu_0 (H + M) \quad (3.1)$$

where μ_0 is the magnetic permeability of vacuum valued at $4\pi \times 10^{-7}$ Wb/(A·m).

Figure 3.1 shows typical M–H and B–H curves where both M and B raise rapidly with the applied field H increasing from zero. With further increased H, the M gradually becomes saturated and approaches a fixed value M_s, known as the saturation magnetization (Figure 3.1a). At this point, the saturation induction B_s is also reached (Figure 3.1b). Although the M is constant after saturation, continuously increased B is observed due to the increment in the H and the slope dB/dH becomes constant beyond the B_s.

DOI: 10.1201/9781003216346-4

After saturation, removing the applied field H does not eliminate the M or B to zero. Instead, they decrease to the residual magnetization M_r and the residual induction B_r, respectively. If the applied field is then reversely increased, the magnetization or induction will be decreased. The reverse field necessary to decrease the M or B to zero corresponds to the coercivity H_c. Sometimes, intrinsic coercivity H_{cj} and normal coercivity H_{cn} are used, corresponding to the reverse field necessary to reduce the M and B to zero, respectively.

With further increased reverse field, magnetization along the opposite direction occurs and saturation shall be achieved at $-M_s$ or $-B_s$. If the reverse field is decreased to zero prior to increasing along the initial direction, the magnetization or induction will pass the point of $-M_r$ or $-B_r$ and reach M_s or B_s again, forming a closed curve as the hysteresis loop.

Such magnetic performance of ferro- and ferrimagnetic materials exists under a critical temperature, namely the Curie temperature T_C, above which the parallel aligned moments transit into random arrangement due to thermal motion, and the ferro- or ferrimagnetism changes into paramagnetism. Figure 3.2 shows typical magnetization curve versus the temperature for ferro- and ferrimagnetic materials, exhibiting rapid decrement in the M approaching T_C.

The magnetic performance of a material is characterized not only by the M or B but also how they vary with the applied field H, which can be reflected by the susceptibility χ as

$$\chi = \frac{M}{H} \tag{3.2}$$

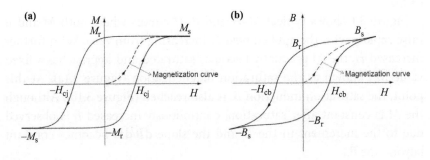

FIGURE 3.1 (a) M–H and (b) B–H curves of typical ferro- or ferrimagnetic materials.

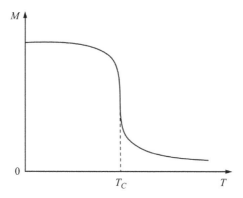

FIGURE 3.2 Typical M–T curves of ferro- or ferrimagnetic materials.

As mentioned in Chapter 2, different types of magnetisms exhibit distinctive feature of the χ. For ferro- or ferrimagnetic materials, the χ is a large positive value, which may change suddenly above the T_C due to the transition into paramagnetism.

The ratio of B to H is named as the permeability μ as follows

$$\mu = \frac{B}{H} \quad (3.3)$$

In fact, relative magnetic permeability $\mu_r = \mu/\mu_0$ is more frequently used. It can be inferred from Equations (3.1) and (3.2) that $\mu_r = 1 + \chi$. The permeability indicates the magnetization sensitivity of a material in the external field. Since the magnetization curve is nonlinear, the μ varies with H and the initial permeability $\mu_i = \frac{1}{\mu_0} \lim_{H \to 0} \frac{B}{H}$ is a specific parameter for performance evaluation.

In practical applications, most of the soft magnetic materials are used under alternating field. At a given frequency, when changing the magnetic field strength, one can get a series of dynamic hysteresis loops as shown in Figure 3.3. The plot connecting the vertices (H_m and B_m) of each dynamic hysteresis loop is called the dynamic magnetization curve.

During dynamic magnetization, various factors including the hysteresis, eddy current, damping and magnetic aftereffect result in time lag phenomenon where the change of B lags behind H with a phase angle of δ. The dynamic field $\widetilde{H}$ and magnetic induction $\widetilde{B}$ are expressed as

$$\widetilde{H} = H_m e^{i\omega t} \quad (3.4)$$

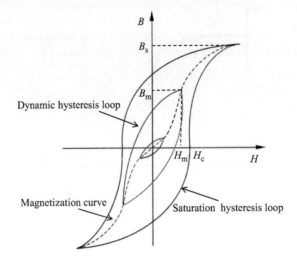

FIGURE 3.3 Dynamic hysteresis loops and magnetization curve.

$$\tilde{B} = B_m e^{i(\omega t - \delta)} \quad (3.5)$$

This gives rise to the complex permeability described by

$$\tilde{\mu} = \frac{\tilde{B}}{\mu_0 \tilde{H}} = \frac{B_m e^{i(\omega t - \delta)}}{\mu_0 H_m e^{i\omega t}} = \frac{B_m}{\mu_0 H_m} \cos\delta - i \frac{B_m}{\mu_0 H_m} \sin\delta$$
$$= \mu_m \cos\delta - i\mu_m \sin\delta = \mu' - i\mu'' \quad (3.6)$$

where the real permeability μ' represents energy storage and the imaginary permeability μ'' reflects energy loss per unit volume when the magnetic materials are magnetized per cycle in the alternating field.

In the complex permeability spectra as shown in Figure 3.4, the critical frequency at which the real part is half of the initial value (or the imaginary part is larger than the real part) is called the cut-off frequency f_r. When $f < f_r$, the μ' decreases and μ'' increases with raised f. Maximum μ'' is reached at $f = f_r$, above which the material cannot work due to the larger loss than the energy storage.

Magnetic loss under the alternating field mainly consists of the hysteresis loss P_h, eddy current loss P_e and residual loss P_r. The P_h is correlated to the area of the hysteresis loop during one magnetization cycle and can be described by $P_h = C_H B^3 f$, while the P_e is due to Joule heating effect

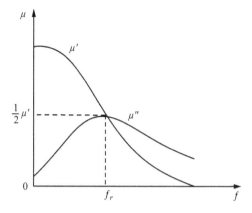

FIGURE 3.4 Typical complex permeability spectra.

resulted from induced eddy current in the alternating field, described by $P_e = C_E B^2 f^2/\rho$ where C_H and C_E are the dimensional constants and ρ is the electrical resistivity (Van Der Zaag, 1999). Power loss except the hysteresis loss and eddy current loss is called the residual loss. It is caused by various magnetic relaxation processes. At low frequencies, the P_r mainly originates in the magnetic aftereffect, whereas at high frequencies, the P_r is mainly derived from dimensional resonance, domain wall resonance and natural resonance, etc. (Périgo et al., 2018).

3.2 MAGNETIC MEASUREMENT TECHNIQUES

Magnetic measurement techniques may be based on different principles including the electromagnetic induction, magneto-optical, magneto-electric or magneto-force effect as well as various resonance phenomena. In the following sections, such techniques are categorized based on their working principles and the function of each technique will be introduced to provide fundamental understanding on the analysis of magnetic structures and evaluation of magnetic performance.

3.2.1 Measurement of Magnetization

Methods for measuring magnetization under static and dynamic fields have been developed usually based on electromagnetic induction. Typical instruments involve vibrating sample magnetometer (VSM), superconducting quantum interference device (SQUID) and B–H analyzer which will be introduced as follows.

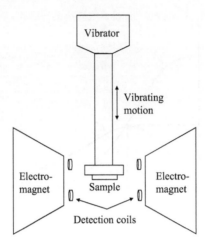

FIGURE 3.5 Schematic Setup of the VSM.

VSM was invented by Simon. Foner at MIT Lincoln Laboratory in 1955 and reported in 1959 (Kronmüller et al., 2007). Figure 3.5 shows the setup of the VSM where a magnetic testing sample is attached to a non-magnetic rod. Vibration of the samples generates changing magnetic flux and induces electromotive force (emf) in the detection coils. The signal is amplified and measured to reflect the magnetic moment of the sample. Standard VSM can detect magnetic moment of around 10^{-5} emu and provides stable measurement in a wide temperature range (Cullity and Graham, 2009). Consequently, not only it can be used to record the hysteresis curve of a material by sweeping the magnetic field, it can also be used for the measurement of the T_C.

SQUID was developed by the Quantum Design company in 1982 using superconducting magnets to excite the magnetic field (Kleiner et al., 2004). Compared with the VSM, much higher sensitivity of ~10^{-7} emu can be achieved for the SQUID (Cullity and Graham, 2009). The SQUID usually consists of a superconducting loop interrupted by either one or two Josephson junctions formed by two weakly coupled superconductors separated by an insulator layer (~1 nm) as in the radio frequency (RF) or direct current (DC) SQUID. Due to the improved sensitivity of the DC SQUID, the RF SQUID was gradually replaced from the 1980s and most of the detections are carried out using the DC SQUID nowadays (Clarke and Braginski, 2004). Figure 3.6 shows typical setup for the DC SQUID where two superconductors are connected end to end to form a ring.

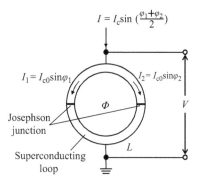

FIGURE 3.6 Schematic illustration of the DC SQUID.

According to the Josephson equations (Josephson, 1962), a supercurrent $I = I_{c0} \sin \varphi_1$ and $I = I_{c0} \sin \varphi_2$ runs through the two tunnel junctions without any applied voltage, where I_{c0} is the critical current, φ_1 and φ_2 are the phase difference between the wave functions of the two superconductors. Under the applied magnetic field, the φ_1 and φ_2 change with the phase modulation of the field based on the flux quantization in the superconducting loop as in (London, 1950)

$$\varphi_1 - \varphi_2 = 2\pi \frac{\Phi}{\Phi_0} \qquad (3.7)$$

where Φ is the applied magnetic flux and $\Phi_0 = h/2e \approx 2.068 \times 10^{-15}$ Wb as the flux quantum. The current passing through the SQUID can then be expressed as (Jaklevic et al., 1964)

$$I = I_1 + I_2 = 2I_{c0} \left| \cos\left(\frac{\Phi}{\Phi_0}\right) \right| \sin \frac{\varphi_1 + \varphi_2}{2} = I_c \sin \frac{\varphi_1 + \varphi_2}{2} \qquad (3.8)$$

Here the current $I_c = 2I_{c0} \left| \cos\left(\frac{\Phi}{\Phi_0}\right) \right|$ is defined as the critical current of the SQUID, which varies with the applied flux with a period of Φ_0. Due to the small Φ_0, slight fluctuation of the external magnetic field would be sufficient to cause distinct change in the I_c. Since it is rather difficult to detect the critical current, the SQUID is biased with a constant current I_b larger than the I_c. In this way, the voltage V across the SQUID is tuned by the magnetic field and oscillates with the Φ_0. As a strong output voltage signal can be generated in response to a small flux change, the SQUID is applied

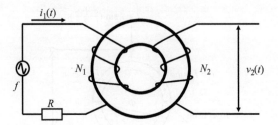

FIGURE 3.7 Schematic illustration of the measurement via B–H analyzer.

to precise measurement of weak magnetic properties especially for biomagnetic and geomagnetic detections due to its ultrahigh sensitivity.

B–H analyzer is utilized to measure dynamic magnetic properties of samples under alternating field. It is also based on electromagnetic induction where an external field is applied to the primary coil and generates an induced voltage on the secondary coil which depends on the magnetic characteristics of the ring sample as illustrated in Figure 3.7. Not only can the dynamic hysteresis loops under varied frequencies be measured using the B–H analyzer, but the effective permeability μ_e and total power loss P_c can also be measured based on

$$\mu_e = \frac{L}{4\pi N_2^2} \times \frac{l_m}{A_m} \times 10^7 \quad (3.9)$$

$$P_c = \frac{N_1}{N_2} \times \frac{1}{T} \times \int_0^T i_1(t) \times v_2(t) \, dt \quad (3.10)$$

where L is the self-inductance, l_m and A_m are the length and cross-sectional area of the magnetic circuit, while N_1, N_2, $i_1(t)$ and $v_2(t)$ are the turn number of primary coil and secondary coil, exciting current and induced voltage, respectively.

3.2.2 Measurements Based on Magnetic-X Effects

3.2.2.1 Measurements Based on Magneto-Optical Effect

Propaganda characteristics of incident light onto magnetic materials may be altered by their magnetic moments as in the Faraday effect and the Kerr effect. The Faraday effect originates in the interaction between the light and atomic magnetic moments which rotates the plane of polarization of light during transmission. Different domains of the specimen may give rise to varied Faraday rotation of the transmitted light for contrast difference in the resultant image (Figure 3.8). Although the Faraday method enables investigation on wall motions, its applications are limited to sufficiently thin samples (up to around 0.1 mm) for light transmission (Cullity and Graham, 2009).

The Kerr effect involves rotated polarization plane for the incident light reflected from the surface of a magnetic sample. The rotation degree θ_K (usually $\ll 1°$) is given by $\theta_K = K_k M$, where K_k is the Kerr constant (Coey, 2010). Similar to the Faraday effect, the Kerr effect is not only used to measure the magnetization, but also for observation of domain walls. The contrast between the adjacent domains, however, tends to be low due to the small θ_K. Thus, high quality and satisfactory alignment are necessary for the optical components. Compared with the measurement on the basis of the Faraday effect, the setup of the optical system here involves the arrangement of the polarizer, specimen, analyzer, and image plane with an angle as illustrated in Figure 3.9. Since it is unnecessary for the incident light to transmit the sample, the Kerr method is applicable to both bulk and thin-film samples.

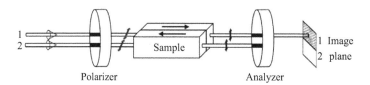

FIGURE 3.8 Magnetic measurement based on the Faraday effect.

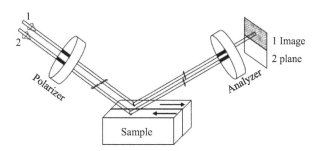

FIGURE 3.9 Magnetic measurement based on the Kerr effect.

3.2.2.2 Measurements Based on Magneto-Electric Effect

When an electric current passes through a solid material, transverse emf may be induced by perpendicularly applied magnetic field or inherent magnetic moments of the material, known as the Hall effect (Hall, 1880). Figure 3.10 demonstrates the Hall effect, where a perpendicular magnetic field H deflects the electrons to accumulate at one side of the sample until the Hall emf is sufficient to balance the Lorentz force. The electric potential of the sample can then be measured to reflect either the applied or intrinsic field strength.

3.2.2.3 Measurements Based on Magneto-Force Effect

When magnetized substances or electrons are put in a non-uniform magnetic field, they will be affected by force along the magnetic field gradient. By measuring the magnetic force of the substance surface or the deflection of transmission electrons, internal magnetic information can be visualized by microscopic imaging technology as in the magnetic force microscopy (MFM) and Lorentz microscopy.

The MFM was invented by Y. Martin and H. K. Wickramasinghe for localized investigation on magnetic properties in 1987 (Martin and Wickramasinghe, 1987). Similar to the atomic force microscopy (AFM), it consists of a tip attached to a cantilever to scan across the sample surface. A major difference, however, is that the tip used in MFM is covered with a ferromagnetic layer with specific magnetization to measure relatively long range (10–100 nm) magnetic tip-sample interaction forces. Images showing the magnetization distribution of the sample surface can be obtained by detecting the cantilever deflection as shown in Figure 3.11. The resolution achievable with MFM is around 20 nm (Zech et al., 2011) which is sufficient for the observation of magnetic microstructures such as magnetic domains and their dynamic magnetization reversal.

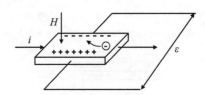

FIGURE 3.10 Hall effect showing the relationship between the field, current and electric potential.

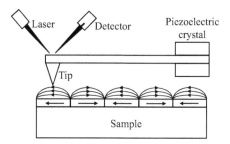

FIGURE 3.11 Schematic drawing of the setup for the MFM.

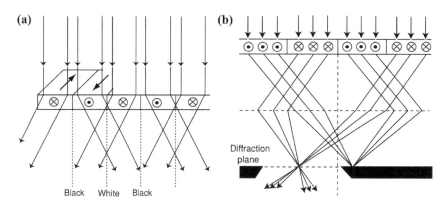

FIGURE 3.12 Fresnel scheme and Foucault scheme of the Lorentz microscopy. Reprinted with permission from (Coey, 2010). Copyright (2010) Cambridge University Press.

The Lorentz microscopy is based on transmission electron microscopic technique for the investigation on the magnetic domain structures. Magnetic samples generate Lorentz force onto the passing electrons, which allows direct observation of interactions between the electrons with the domain walls, crystal imperfections and grain boundaries with high resolution. During the operation of the Lorentz microscope, two modes are available, including the Fresnel scheme and the Foucault scheme as illustrated in Figure 3.12 (Coey, 2010). In the Fresnel mode, defocused imaging lens are used to generate dark and bright regions for different domains in an otherwise contrast-free image. It is, however, not possible to obtain any information on the magnetization direction of the domain in this mode. For the Foucault mode, a contrast forming aperture is used at the diffraction plane, so that the magnetization orientation determines whether electrons could pass through the aperture to form bright and dark regions in the resultant image.

3.2.2.4 Measurements Based on Magnetic Resonance

The application of an alternating magnetic field onto a substance may induce resonances, either involving magnetic moments of the electrons as in the ferromagnetic resonance (FMR) or those of the nucleus as in the Mössbauer spectroscopy. The FMR is based on electron spin resonance using a signal source to emit microwaves. When the frequency of the microwaves is coincident with the magnetization precession frequency of the sample, FMR occurs for significantly increased microwave absorption. The FMR reflects uniform resonance of electron spin, the signal of which is sufficiently strong to detect small mass of the ferromagnetic samples (2–10 mg) (Cullity and Graham, 2009). It should be noted that the FMR curves may also contain multiple absorption peaks caused by various non-uniform resonance modes such as domain wall resonance.

The Mössbauer effect is based on the nuclear spin resonance discovered by R. L. Mössbauer (Mössbauer, 1958). During the measurement, gamma rays are incident on the sample surface and their intensity after transmitting through the sample is recorded. If the velocity of the gamma ray is equivalent to the resonant energy level of the sample, a fraction of the gamma ray is absorbed, leading to a drop in the resultant intensity spectrum. Consequently, chemical environment of the nuclei under measurement can be revealed based on careful analysis of the location and intensity of the dips.

The above-introduced techniques have been widely applied to investigate magnetic properties from macroscopic quantity to microscopic structure. The measurement techniques are, however, by no means exhaustive. Apart from experimental techniques, theoretical calculations and simulations have also advanced our understandings on magnetism and magnetic materials. For instance, ab initio has been used to calculate spin and distribution of magnetic moment. Micromagnetic simulation provides additional method to study local magnetic properties at the nano/microscale. Electromagnetic simulation packages based on finite element method such as COMSOL, CST and HFSS software have also been adopted to investigate the distribution of magnetic field. To understand a specific case, combined magnetic measurement techniques and theoretical calculation tools may be necessary to provide an all-around picture.

REFERENCES

Clarke, J., and A. I. Braginski. 2004. *The SQUID handbook: fundamentals and technology of SQUIDs and SQUID systems.* Weinheim: Wiley-VCH.

Coey, J. M. D. 2010. *Magnetism and magnetic materials.* New York: Cambridge University Press.

Cullity, B. D., and C. D. Graham. 2009. *Introduction to magnetic materials.* Piscataway: Wiley-IEEE Press.

Hall, E. H. 1880. On a new action of the magnet on electric currents. *American Journal of Science* 111:200–5.

Jaklevic, R. C., J. Lambe, A. H. Silver, and J. E. Mercereau. 1964. Quantum interference effects in Josephson tunneling. *Physical Review Letters* 12:159–60.

Josephson, B. D. 1962. Possible new effects in superconductive tunnelling. *Physics Letters* 1:251–3.

Kleiner, R., D. Koelle, F. Ludwig, and J. Clarke. 2004. Superconducting quantum interference devices: state of the art and applications. *Proceedings of the IEEE* 92:1534–48.

Kronmüller, H., S. Parkin, R. Waser, U. Böttger, and S. Tiedke. 2007. *Handbook of magnetism and advanced magnetic materials.* Hoboken: John Wiley & Sons.

London, F. 1950. *Superfluids.* New York: John Wiley & Sons.

Martin, Y., and H. K. Wickramasinghe. 1987. Magnetic imaging by force microscopy with 1000 Å resolution. *Applied Physics Letters* 50:1455–7.

Mössbauer, R. L. 1958. Kernresonanzfluoreszenz von gammastrahlung in Ir^{191}. *Zeitschrift für Physik* 151:124–43.

Périgo, E. A., B. Weidenfeller, P. Kollár, and J. Füzer. 2018. Past, present, and future of soft magnetic composites. *Applied Physics Reviews* 5:031301.

Zaag, P. J. Van Der. 1999. New views on the dissipation in soft magnetic ferrites. *Journal of Magnetism and Magnetic Materials* 196–197:315–9.

Zech, M., C. Boedefeld, F. Otto, and D. Andres. 2011. Magnetic imaging on the nanometer scale using low-temperature scanning probe techniques. *Microscopy Today* 19:34–8.

FURTHER READING

Tumanski, S. 2011. *Handbook of magnetic measurements.* Boca Raton: CRC Press Inc.

II

Hard Magnetic Materials

CHAPTER 4

Introduction to Hard Magnetic Materials

4.1 PERFORMANCE REQUIREMENTS FOR HARD MAGNETIC MATERIALS

4.1.1 Intrinsic Magnetic Properties

From the basic physical perspective, understanding the performance of hard magnetic materials should begin with the fundamental behavior of ferromagnetic materials. An important feature is the spontaneous alignment of the magnetic moments, which arise predominantly from the contribution of unpaired electrons (Franse and Radwański, 1993; Harris and Jewell, 2012). As represented in Figure 4.1a, under an equilibrium condition with minimized total magnetic energy (mainly magnetostatic and exchange energy), the formed magnetic domains exhibit different orientation directions (Herbst, 1991). As a result, the total magnetization moments are averaged to zero, hence the bulk material does not exhibit any external magnetic flux. When applying an increasing external field, the magnetic domains gradually align to the direction of the applied field until the saturation magnetization M_s is reached below the Curie temperature T_C (Figure 4.1b). For the ferromagnetic transitional metals (TM, such as Fe, Co and Ni), the M_s value in proportion to the numbers of unpaired 3d electrons is lowered from Fe ($3d^6$) to Co ($3d^7$) to Ni ($3d^8$) (Burzo, 1998). At elevated temperatures, the M_s value degrades due to the progressive

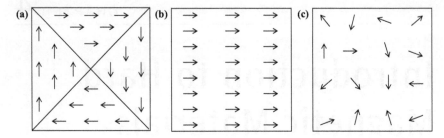

FIGURE 4.1 (a) Alignment of magnetic moments without application of external field below T_C, as well as those with applied field (b) below and (c) above T_C for ferromagnetic materials.

enhanced thermal perturbation, and the ferromagnetism transforms to paramagnetism above T_C, as depicted in Figure 4.1c. In addition to the M_s and T_C, the magnetocrystalline anisotropy field H_A is another intrinsic property of hard magnetic materials. For the TM, low H_A is a consequence of relatively weak coupling between spin moment μ_s and orbital moment μ_l of 3d electron (Hirosawa et al., 1986), particularly given the fact that the outermost 3d electrons confined by the crystal-field interactions usually exhibit quenched orbital angular moment, as discussed in Chapter 2. However, the rare earth (RE) metals with stronger 4f spin-orbit coupling generate higher H_A (Kanamori, 2006). As a result, through the combination of TM and RE metals, RE-based hard magnetic materials with synergistically high M_s and H_A have been developed, represented by the first-generation SmCo$_5$, the second-generation Sm$_2$Co$_{17}$ and the third-generation Nd$_2$Fe$_{14}$B.

4.1.2 Hysteresis Loop

Figure 4.2 shows the broad M–H and B–H hysteresis loops of typical hard magnetic material. The basic requirements for hard magnetic material involve large remanence B_r, intrinsic coercivity H_{cj} and maximum energy product $(BH)_{max}$, which are extrinsic properties as illustrated by the second-quadrant demagnetization curve. From the initial magnetization curve, the domain structures evolve upon the increasingly applied magnetic field H. Inset (a) shows the initial ferromagnetic domains separated by the domain walls, which are randomly aligned in the unmagnetized state. Insets (b) and (c) represent that the magnetic domains gradually align parallel to the applied field until reaching the M_s value. Note that the dotted curves (i) and (ii) represent the different magnetization

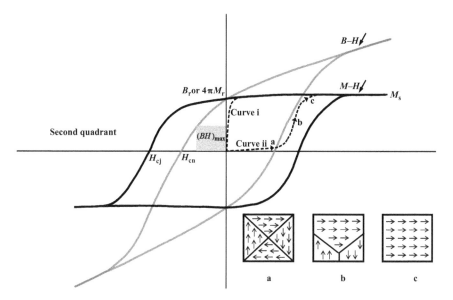

FIGURE 4.2 Representative hysteresis curves of hard magnetic materials.

mechanism, which will be described in Section 4.1.4. With reduced magnetizing field H back to zero, the remaining magnetization is defined as remanence B_r (or $4\pi M_r$). Furthering increasing the reversal field, the reversal domains appear, and the magnetization decreases gradually. As mentioned in Chapter 3, the magnetic field at which the magnetization M or flux density B is reduced back to zero corresponds to the intrinsic coercivity H_{cj} (from the M-H loop) or normal coercivity H_{cn} (from the B-H loop). The energy product (BH) is defined as the product between B and H in the second-quadrant B-H curve, and its maximum value is called the maximum energy product $(BH)_{max}$, being an important parameter to evaluate the performance of hard magnetic materials.

4.1.3 Remanence

Remanence B_r refers to the remaining flux density after removing the external magnetic field ($H = 0$), which is an important indicator of the flux-producing capability of a given hard material. The B_r can be expressed in the unit of T (SI system) or Gs (CGS system), following the relationship of 1 T = 10^4 Gs. To achieve a large B_r, microstructural characteristics including high degree of crystalline orientation toward the external field direction, minimum fraction of impurity phase and full densification are preferred. Taking sintered NdFeB magnet as an example, B_r can be simply calculated according to (Zhou and Dong, 2004)

$$B_r = A(1-\beta)\frac{d_{\text{experimental}}}{d_{\text{theoretical}}}\overline{\cos\theta}J_s \quad (4.1)$$

Here, A denotes the volume fraction of positive domains. Due to the misalignment of magnetic domains caused by soft defects and demagnetization field in the NdFeB magnet, reversal domains tend to exist at the remanent state (Hono and Sepehri-Amin, 2018). β represents the volume fraction of impurity phase, such as RE-oxide, RE-rich and B-rich phases that are detrimental to the remanence. $\frac{d_{\text{experimental}}}{d_{\text{theoretical}}}$ is the relative density resulted from incomplete densification. $\overline{\cos\theta}$ denotes the alignment degree of matrix phase grains, with θ related to the deviation angle of the easy axis from the alignment axis. J_s is the intrinsic saturation magnetic polarization of the 2:14:1 matrix phase. Accordingly, basic approaches to enhance the remanence of sintered NdFeB magnets can be summarized as: (i) appropriate reduction of the RE content during composition design to enlarge the volume fraction of 2:14:1 matrix phase; (ii) low-oxygen process to suppress the formation of RE oxides; (iii) regulating the magnetic orientation and compaction procedures to obtain a high alignment degree along the easy c-axis; (iv) optimizing the sintering temperature and time to achieve full densification. As displayed in Figure 4.3, a close-to-ideal

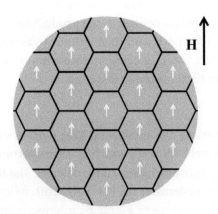

FIGURE 4.3 Schematic of close-to-ideal microstructure toward high-remanence NdFeB permanent magnet.

microstructure of the NdFeB permanent magnet involves a small amount of thin Nd-rich phase layers finely dispersed to fully dense the 2:14:1 crystallites with consistent c-axis orientation.

4.1.4 Coercivity

Intrinsic coercivity H_{cj} represents the resistance against field or thermal demagnetization of the hard magnetic materials. Theoretically, the upper limit of coercivity is determined by the magnetocrystalline anisotropy H_A as in

$$H_{cj}^{max} = H_A = 2K_1 / \mu_0 M_s \tag{4.2}$$

where K_1 denotes the constant of magnetocrystalline anisotropy (assuming $K_1 \gg K_2$), μ_0 denotes the permeability of free space. The H_{cj} can be expressed in the unit of A/m (SI system) or Oe (CGS system), following the relationship of 1 Oe = $10^3/4\pi$ A/m = 79.6 A/m.

To maximize the coercivity, the average grain size D of ferromagnetic phase should be refined to the single-domain regime, where the magnetization reversal occurs through coherent rotation of the magnetic vector from one easy axis to another against the magnetocrystalline anisotropy (Mohapatra and Liu, 2018). With significantly larger D than the critical single-domain size D_c (~300, 1000 and 2000 nm for $Nd_2Fe_{14}B$, hard ferrites and $SmCo_5$, respectively), the magnetic structure splits into magnetic domains, and magnetization occurs by domain wall motion, causing the coercivity degradation. If the grain size is below D_c, the superparamagnetism occurs as introduced in Chapter 2, upon which the coercivity becomes negligible. In reality, the practical coercivity H_{cj} is usually 20–30% of the theoretical H_A even within the single-domain regime, which is acknowledged as the Brown's paradox (Brown, 1945). Such a huge discrepancy is generally attributed to the random occurrence of crystallographic defects, surface imperfections and so on. According to Kronmüller and coworker's approximation (Kronmüller et al., 1988), a general expression of coercivity H_{cj} can be defined by

$$H_{cj} = \alpha H_A - N_{eff} M_s \tag{4.3}$$

where α ($0 < \alpha < 1$) is associated with the imperfections with reduced anisotropy field and misaligned grains, and N_{eff} is defined as the effective demagnetizing factor.

Besides the coherent rotation of single-domain permanent magnets, both the reversal domain nucleation and domain wall pinning are also able to dominate the coercivity of permanent magnets. For the nucleation model, the H_{cj} is determined by the nucleation field H_N of reversal domain, and the M_s can be reached rapidly when applying a relatively low magnetizing field (indicated by the dotted magnetization curve i in Figure 4.2) (Xu et al., 2020; Kim et al., 2019; Sasaki et al., 2016). Typical examples are SmCo$_5$ and sintered NdFeB magnets. Hence, extensive studies have been carried out to minimize the initial deficient sites for reversal domain nucleation (Liu et al., 2020; Tang et al., 2021). Main approaches involve the construction of continuous grain boundary (GB) phase to decouple the neighboring ferromagnetic grain, and magnetically hardening shell to enhance local anisotropy field. For the pinning model, the structural or magnetic inhomogeneities behave as pinning sites to impede the domain wall motion, which increases the difficulty of magnetizing the hard magnetic material with low-field strength (indicated by the dotted curve ii in Figure 4.2) (Wu et al., 2020; Song et al., 2021). For the pinning-type Sm$_2$Co$_{17}$ magnet, increasing the density of pinning sites and enhancing the pinning strength of 1:5 cell boundaries are common approaches to impede the domain wall motion during magnetization reversal (Chen et al., 2019; Duerrschnabel et al., 2017).

4.1.5 Energy Product

Energy product (BH) can be obtained for each individual point in the second-quadrant B-H hysteresis curve, among which a maximum value $(BH)_{max}$ exists as shown in Figure 4.4. Since the hard magnetic material

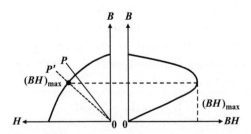

FIGURE 4.4 Demagnetization curve in the second quadrant of B-H hysteresis curve (left), and the corresponding values of energy product (BH) plotted on the same B scale (right). The $(BH)_{max}$ is indicated in both plots.

usually operates under an open-circuit state, the resultant free poles shall generate a demagnetizing field H_d, leading to a lower induction than the B_r in a closed ring. The operating point P is highly dependent on the slope of the load line, as determined by the magnet geometry and demagnetizing factor N_d between 0 and 1. Usually, the N_d takes 0 for long needles, 1/3 for spheres, and 1 for thin films with perpendicular anisotropy, which makes the thin magnet difficult for stable operation. Evidently, in order to shift the operation point from P to P' for the magnet of Figure 4.4 toward the $(BH)_{max}$ point, the magnet should be designed in a thicker or shorter shape.

For hard magnetic materials, the $(BH)_{max}$ is deemed as a critical figure of merit quantifying the stored energy per unit volume. It combines two key parameters, remanence B_r and coercivity H_{cj} representing the magnetic strength and demagnetizing resistance, respectively. Evidently, the higher the $(BH)_{max}$, the lower volume of the hard magnetic material is required for given magnetic energy in the finished device. The $(BH)_{max}$ can be expressed in MGOe (CGS system) or kJ/m³ (SI system), following the relationship of 1 MGOe = $10^2/4\pi$ kJ/m³ = 7.96 kJ/m³. The past century has witnessed the successive breakthroughs of the $(BH)_{max}$, spanning from ~1 MGOe for steels, to ~30 MGOe for Sm_2Co_{17} magnets, and further to ~56 MGOe for $Nd_2Fe_{14}B$ magnets (Gutfleisch et al., 2011).

The upper limit on the energy product is $(1/4)\mu_0 B_r^2$, with the squareness factor of 1.0 (determined by the ratio of J at 10% H_{cj} to J_r) (Branagan et al., 1999). However, due to the non-ideal shape of actual hysteresis loop affected by the grain size distribution, grain shape and surface morphology, the practical energy product falls far from this value. Since the $(BH)_{max}$ is closely related to the B_r, any factor that is beneficial for enlarged B_r shall give rise to the enhanced $(BH)_{max}$.

4.1.6 Temperature Coefficients

Stability over a range of the recommended service temperatures is an important factor to be considered for the hard magnetic materials. Typically, the permanent magnets in hybrid vehicle generators are required routinely to perform above 150 °C, requiring a strong resistance to the thermal demagnetization. However, on most occasions, both the B_r and H_{cj} decline naturally with increased operating temperature. Such irreversible losses are usually evaluated by two metrics, including the temperature coefficient of remanence α and temperature coefficient of coercivity β, which are expressed as follows (Brown et al., 2002)

$$\alpha = \frac{B_r(T_2) - B_r(T_1)}{B_r(T_1)(T_2 - T_1)} \times 100\% \qquad (4.4)$$

$$\beta = \frac{H_{cj}(T_2) - H_{cj}(T_1)}{H_{cj}(T_1)(T_2 - T_1)} \times 100\% \qquad (4.5)$$

where $B_r(T_1)$, $B_r(T_2)$, $H_{cj}(T_1)$ and $H_{cj}(T_2)$ denote the remanence and coercivity of the magnets at temperature T_1 and T_2, respectively. In the following chapters, the hard magnetic materials with positive or negative thermal coefficients will be discussed, which are tailored for different industrial applications.

4.2 DEVELOPMENT OF HARD MAGNETIC MATERIALS

Figure 4.5 plots the progresses of $(BH)_{max}$ values versus time during the past century (Gutfleisch et al., 2011). In the 1910s, the steels containing carbon, tungsten, cobalt or chromium have been developed as the early permanent magnets. Although the steels possess rather large B_r of ~10 kG, their coercivity is limited to the orders of a few hundred Oe, which are no longer used as permanent magnets due to the development of other alternatives with higher $(BH)_{max}$. On the basis of primary steel, a next class of Alnico has been developed in the 1930s, which was once the most widely used permanent magnet due to its high T_C and satisfactory thermal stability, but now has been largely replaced by the hard ferrites and RE-based permanent magnets (Cui et al., 2018; Sugimoto, 2011). Note that the coercivity of Alnico arises from the shape anisotropy that a rod-shape ferromagnetic phase uniformly disperses within the weakly magnetic matrix. In 1952, Went et al. first reported a relatively large uniaxial crystal anisotropy for the hexagonal barium ferrite $BaFe_{12}O_{19}$ (Went et al., 1952). Later on, Cochardt discovered the strontium ferrite $SrFe_{12}O_{19}$ with higher anisotropy and lower density, exhibiting a $(BH)_{max}$ of 5 MGOe (Cochardt, 1963). Although the barium and strontium ferrites exhibit much lower M_s compared with the metallic hard materials, they are the choice of low-to-medium grade applications in loudspeakers and automotive devices.

The following three decades witnessed a new era of RE-based permanent magnets, generating the significantly enlarged $(BH)_{max}$ exceeding any

Introduction to Hard Magnetic Materials ■ 53

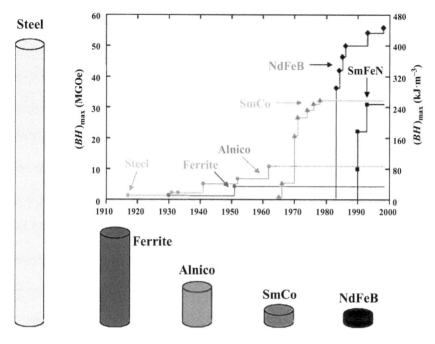

FIGURE 4.5 Significantly improved $(BH)_{max}$ and device miniaturization pushed by the development of permanent magnets in the 20th century. Reprinted with permission from (Gutfleisch et al., 2011). Copyright (2011) John Wiley and Sons.

previously attained values. The pure RE elements alone with strong magnetic anisotropy are limited by their low T_C (below room temperature). When the REs are integrated with TMs to form the intermetallic compounds, both large T_C and M_s can be achieved simultaneously. Motivated by this idea, massive search has begun for the RE-TM alloys, with the RE component providing large magnetocrystalline anisotropy intrinsically responsible for the coercivity, and the TM component principally affording high magnetization. Hexagonal $CaCu_5$-type $SmCo_5$ compound as the first-generation RE-based permanent magnet was invented in the 1960s. Strnat et al. reported $(BH)_{max}$ of 5.1 MGOe for the $SmCo_5$ bonded magnet (Strnat et al., 1967). Shortly after, Das fabricated the $SmCo_5$ sintered magnets with $(BH)_{max}$ of 20 MGOe via powder metallurgy (Das, 1969). With higher Co content and accordingly larger M_s, $(BH)_{max}$ and T_C than those of $SmCo_5$, Sm_2Co_{17} has later become a focus to permit high-temperature operation (Narasimhan and Wallace, 1977). Ojima et al. from TDK Company prepared the Sm_2Co_{17} sintered magnets to deliver a large $(BH)_{max}$

up to 30 MGOe, representing the birth of second-generation RE-based permanent magnet (Ojima et al., 1977).

Although $SmCo_5$ and Sm_2Co_{17} exhibit theoretical $(BH)_{max}$ above ~30 MGOe as well as satisfactory T_C above ~750 °C, the consumption of the strategic Co and the least abundant Sm limits their further development. Such economic constraints, in-tune have spurred extensive investigations to find new alternatives with lower cost and better magnetic performance, which eventually led to the development of NdFeB as the third-generation RE-based permanent magnet. In 1984, Sagawa et al. from Sumitomo Special Metals Co. and Croat et al. from General Motors Corp. independently discovered the $Nd_2Fe_{14}B$-based permanent magnets through two different techniques, liquid-phase-sintering and melt-spinning (Sagawa et al., 1984; Croat et al., 1984). NdFeB magnets exhibit great economic advantages since they do not contain any strategic metal Co and the reserve of Nd is 5-10 times richer than that of Sm (Yan and Peng, 2019). With the superior intrinsic magnetic properties of $Nd_2Fe_{14}B$ tetragonal phase (J_s = 1.60 T, T_C = 312 °C, H_A = 7.3 T), and theoretical $(BH)_{max}$ as high as 64 MGOe, NdFeB has rapidly replaced SmCo magnets. The NdFeB magnets fabricated in laboratories exhibit a record high $(BH)_{max}$ of 59 MGOe, and even commercial NdFeB with a $(BH)_{max}$ as high as 54 MGOe can be massively produced. The search for novel permanent magnets continues after NdFeB, with the emergence of exchange-coupled nanocomposites, $Sm_2Fe_{17}N_x$ interstitial magnet, and non-RE candidates such as MnBi, MnAl and MnAlC, which will be further discussed in Chapter 8.

4.3 APPLICATIONS OF HARD MAGNETIC MATERIALS

Figure 4.6 provides a first view of the global market for hard magnetic materials (Coey 2012). Hard ferrites with $(BH)_{max}$ ≤ 5.5 MGOe and NdFeB with $(BH)_{max}$ ≥ 30 MGOe are produced on a large scale, being the currently dominant hard magnetic materials. Hard ferrites represent about one-third of the permanent magnet market. Most of the remaining two-thirds are based on NdFeB, among which over 80% are produced in China, the main RE reserve. Around 5% of the RE-based magnets belong to the SmCo family, which are intended for high-temperature applications. There also remains a small but steady demand (1–2%) for the traditional Alnico magnets.

In our contemporary society, the global energy transition from fossil fuels to environmentally friendly renewables, places the hard magnetic

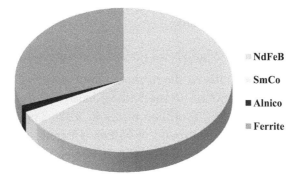

FIGURE 4.6 Breakdown of the market for hard magnetic materials. Reprinted with permission from (Coey 2012). Copyright (2012) Elsevier.

materials at the forefront of modern technologies. In particular, the hard magnetic materials find technologically important and advanced applications in wind power generation, new energy vehicle, information technology and industrial robots, as discussed in the following sections.

4.3.1 Hard Magnetic Materials for Wind Power Generation

Compared with traditional fossil energy, wind energy provides a clean and stable renewable resource with worldwide applicability. For such energy conversion, generators are essential. Different from traditional asynchronous generators using soft magnetic lamination and armature windings, the direct-drive synchronous generators based on permanent magnets are superior with high efficiency, long lifespan, high reliability and simple structure. It is estimated that a 3 MW direct-drive generator consumes ~1.5 tons of NdFeB permanent magnets. The cumulative installed capacity in 2050 may reach 2500 GW, indicating the consumption of 1–2 million tons of NdFeB in the wind power industry.

4.3.2 Hard Magnetic Materials for New Energy Vehicle

Diversified new energy vehicles (electric vehicles, fuel cell vehicles, hybrid electric vehicles, etc.) on the rise to replace the traditional gasoline-powered combustion engines suggest continual increasing demand for permanent magnets. In 2018, the global sales of new energy vehicles exceeded 2 million, consuming nearly 4000 tons of NdFeB sintered magnets (Yan and Jin, 2020). In the future, the trend of miniaturization and lightweight in automobile manufacturing also urges the development of high-performance hard magnetic materials.

4.3.3 Hard Magnetic Materials for Information Technology

For the information industry, the impact of flourishing solid state drives results in a declining output of mechanical hard disk drives, from 630 million pieces in 2010 to 370 million pieces in 2018 (Yan and Jin, 2020). Correspondingly, the consumption of NdFeB sintered magnets in voice coil motors has dropped from 6260 tons to 3700 tons, and NdFeB bonded magnets in hard disk spindle motors has decreased from 1500 tons to 1110 tons. However, toward the new era of "Internet +" with tremendous data, there is a high prospect of cloud computing and 5G network in the future. In this context, the demand for hard disk drives with storage capacity above 1 TB is expected to experience explosive growth in the medium-to-long term, generating a new boom of permanent magnet industry.

4.3.4 Hard Magnetic Materials for Industrial Robot

Nowadays, industrial robotics has been recognized as the key enabler of Industry 4.0 and as the most representative intelligent manufacturing industry. The prevalence of industrial robotics holds particularly for China with a strong automotive sector, with the sales peaking at ~148,000 units in 2018 that occupies one-third of the global market share (Yan and Jin, 2020). This places extremely high demands on the motors and sensors, where the permanent magnets, especially the ultra-small RE-based permanent magnet servomotors with high energy density are indispensable components. To realize the purpose of automatic control and multipurpose manipulation in three or more axes, large amounts of high-performance servomotors are required. For example, the facial expression alone requires nearly 30 pieces of RE-based permanent magnet servomotors.

In summary, the hard magnetic materials with large B_r, H_{cj} and $(BH)_{max}$ are essential for numerous applications requiring fixed sources of magnetic flux, particularly for the emerging new-energy applications represented by the wind power generators and electric motors. Depending on the different composition and magnetic features, the hard magnetic materials can be divided into the RE-free and RE-based ones. Details of each category of the hard magnetic materials that are important to the practical applications will be provided in the following chapters.

REFERENCES

Branagan, D. J., M. J. Kramer, Y. L. Tang., and R. W. McCallum. 1999. Maximizing loop squareness by minimizing gradients in the microstructure. *Journal of Applied Physics* 85(8):5923–5.

Brown, D., B. M. Ma, and Z. M. Chen. 2002. Developments in the processing and properties of NdFeB-type permanent magnets. *Journal of Magnetism and Magnetic Materials* 248(3):432–40.

Brown, W. F. 1945. Virtues and weaknesses of the domain concept. *Review of Modern Physics* 17(1):15–9.

Burzo, E. 1998. Permanent magnets based on R-Fe-B and R-Fe-C alloys. *Reports on Progress in Physics* 61(9):1099–266.

Chen, H. S., Y. Q. Wang, Y. Yao, J. T. Qu, F. Yun, Y. Q. Li, S. P. Ringer, M. Yue, and R. K. Zheng. 2019. Attractive-domain-wall-pinning controlled Sm-Co magnets overcome the coercivity-remanence trade-off. *Acta Materialia* 164:196–206.

Cochardt, A. 1963. Modified strontium ferrite, a new permanent magnet material. *Journal of Applied Physics* 34(4):1273–4.

Coey, J. M. D. 2012. Permanent magnets: plugging the gap. *Scripta Materialia* 67(6):524–9.

Croat, J. J., J. F. Herbst, R. W. Lee, and F. E. Pinkerton. 1984. Pr-Fe and Nd-Fe-based materials: a new class of high-performance permanent magnets. *Journal of Applied Physics* 55(6):2078–82.

Cui, J., M. Kramer, L. Zhou, F. Liu, A. Gabay, G. Hadjipanayis, B. Balasubramanian, and D. Sellmyer. 2018. Current progress and future challenges in rare-earth-free permanent magnets. *Acta Materialia* 158:118–37.

Das, D. K. 1969. Twenty million energy product samarium-cobalt magnet. *IEEE Transactions on Magnetics* 5(3):214–6.

Duerrschnabel, M., M. Yi, K. Uestuener, M. Liesegang, M. Katter, H. J. Kleebe, B. Xu, O. Gutfleisch, and L. Molina-Lunal. 2017. Atomic structure and domain wall pinning in samarium-cobalt-based permanent magnets. *Nature Communications* 8:54.

Franse, J. J. M., and R. J. Radwański. 1993. Magnetic properties of binary rare-earth 3d-transition-metal intermetallic compounds. In *Handbook of magnetic materials*, ed. E. Brück, 7:307–501. Amsterdam: North-Holland Publishing.

Gutfleisch, O., M. A. Willard, E. Brück, C. H. Chen, S. G. Sankar, and J. P. Liu. 2011. Magnetic materials and devices for the 21st Century: stronger, lighter, and more energy efficient. *Advanced Materials* 23:821–42.

Harris, I. R., and G. W. Jewell. 2012. Rare-earth magnets: properties, processing and applications. In *Functional materials for sustainable energy applications*, ed. J. A. Kilner, S. J. Skinner, S. J. C. Irvine, and P. P. Edwards, 600–39. Cambridge: Woodhead Publishing.

Herbst, J. F. 1991. $R_2Fe_{14}B$ materials: intrinsic properties and technological aspects. *Reviews of Modern Physics* 63(4):819–98.

Hirosawa, S., Y. Matsuura, H. Yamamoto, S. Fujimura, M. Sagawa, and H. Yamauchi. 1986. Magnetization and magnetic anisotropy of $R_2Fe_{14}B$ measured on single crystals. *Journal of Applied Physics* 59(3):873–9.

Hono, K., and H. Sepehri-Amin. 2018. Prospect for HRE-free high coercivity Nd-Fe-B permanent magnets. *Scripta Materialia* 151:6–13.

Kanamori, J. 2006. Rare earth elements and magnetism in metallic systems. *Journal of Alloys and Compounds* 408–412:2–8.

Kim, T. H., T. T. Sasaki, T. Ohkubo, Y. Takada, A. Kato, Y. Kaneko, and K. Hono. 2019. Microstructure and coercivity of grain boundary diffusion processed Dy-free and Dy-containing Nd-Fe-B sintered magnets. *Acta Materialia* 172:139–49.

Kronmüller, H., K. D. Durst, and M. Sagawa. 1988. Analysis of the magnetic hardening mechanism in RE-Fe-B permanent magnets. *Journal of Magnetism and Magnetic Materials* 74(3):291–302.

Liu, Y. S., J. Y. Jin, M. Yan, M. X. Li, B. X. Peng, Z. H. Zhang, and X. H. Wang. 2020. A reliable route for relieving the constraints of multi-main-phase Nd-La-Ce-Fe-B sintered magnets at high La-Ce substitution: (Pr, Nd)H_x grain boundary diffusion. *Scripta Materialia* 185:122–8.

Mohapatra, J., and J. P. Liu. 2018. Rare-earth-free permanent magnets: the past and future. In *Handbook of magnetic materials*, ed. E. Brück, 27:1–57. Amsterdam: North-Holland Publishing.

Narasimhan, K., and W. E. Wallace. 1977. Magnetic-anisotropy of substituted R_2Co_{17} compounds (R=Nd, Sm, Er, and Yb). *IEEE Transactions on Magnetics* 13(5):1333–5.

Ojima, T., S. Tomizawa, T. Yoneyama, and T. Hori. 1977. Magnetic properties of a new type of rare-earth cobalt magnets $Sm_2(Co, Cu, Fe, M)_{17}$. *IEEE Transactions on Magnetics* 13(5):1317–9.

Sagawa, M., S. Fujimura, N. Togawa, H. Yamamoto, and Y. Matsuura. 1984. New material for permanent magnets on a base of Nd and Fe. *Journal of Applied Physics* 55(6):2083–7.

Sasaki, T. T., T. Ohkubo, and K. Hono. 2016. Structure and chemical compositions of the grain boundary phase in Nd-Fe-B sintered magnets. *Acta Materialia* 115:269–77.

Song, X., T. Y. Ma, X. L. Zhou, F. Ye, T. Yuan, J. D. Wang, M. Yue, F. Liu, and X. B. Ren. 2021. Atomic scale understanding of the defects process in concurrent recrystallization and precipitation of Sm-Co-Fe-Cu-Zr alloys. *Acta Materialia* 202:290–301.

Strnat, K., G. Hoffer, J. Olson, W. Ostertag, and J. J. Becker. 1967. A family of new cobalt-base permanent magnet materials. *Journal of Applied Physics* 38(3):1001–2.

Sugimoto, S. 2011. Current status and recent topics of rare-earth permanent magnets. *Journal of Physics D: Applied Physics* 44(6):064001.

Tang, X., J. Li, H. Sepehri-Amin, T. Ohkubo, K. Hioki, A. Hattori, and K. Hono. 2021. Improved coercivity and squareness in bulk hot-deformed Nd-Fe-B magnets by two-step eutectic grain boundary diffusion process. *Acta Materialia* 203:116479.

Went, J. J., G. W. Rathenau, and E. W. Gorter. 1952. A class of permanent magnetic materials. *Philips Technical Review* 13:194–208.

Wu, H. C., C. Y. Zhang, Z. Liu, G. Q. Wang, H. M. Lu, G. X. Chen, Y. Li, R. J. Chen, and A. R. Yan. 2020. Nanoscale short-range ordering induced cellular structure and microchemistry evolution in Sm_2Co_{17}-type magnets. *Acta Materialia* 200:883–92.

Xu, X. C., Y. Q. Li, Z. H. Ma, M. Yue, and D. T. Zhang. 2020. Sm_2Co_7 nanophase inducing low-temperature hot deformation to fabricate high performance $SmCo_5$ magnet. *Scripta Materialia* 178:34–8.

Yan, M., and X. L. Peng. 2019. *Fundamentals of magnetics and magnetic materials.* HangZhou: Zhejiang University Press.

Yan, M., and J. Y. Jin. 2020. New material industry: rare earth based magnetic materials. In *2020 report on the development of China's strategic emerging industries.* Beijing: Science Press.

Zhou, S. Z., and Q. F. Dong. 2004. *Super permanent magnet: rare earth-iron based permanent magnetic material.* Beijing: Metallurgical Industry Press.

FURTHER READING

Coey, J. M. D. 2011. Hard magnetic materials: a perspective. *IEEE Transactions on Magnetics* 47:4671–81.

CHAPTER 5

Rare-Earth-Free Hard Magnetic Materials

5.1 HARD FERRITES

5.1.1 Introduction

The hard ferrite possesses relatively high coercivity, and is an important family of hard magnetic materials. Since the first release by Philips Company (Netherlands) in 1952 (Went et al., 1952), the $MFe_{12}O_{19}$ (divalent M = Ba, Sr, Pb) with spindle-type hexagonal structure has become the most widely used commercial hard ferrite (Sugimoto, 1999; McCallum et al., 2014). Although successive advents of novel permanent materials particularly the RE-based hard magnetic materials with overwhelmingly superior $(BH)_{max}$ have greatly propelled the enhancement of magnetic properties, the M-type hard ferrites as the least expensive magnets are still massively manufactured and applied in a multitude of consumer electrics and automotive devices. As displayed in Figure 5.1, the annual sales of hard ferrites in China continually grows from $1.6 billion in 2012 to $2.2 billion in 2019 (statistic data provided by China Electronic Components Association). Given the global annual production of hard ferrites at one million tons, it averages out at ~132 g for everyone on earth.

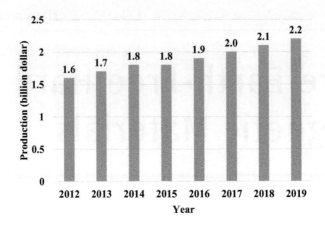

FIGURE 5.1 Output of hard ferrites in China during 2012–2019.

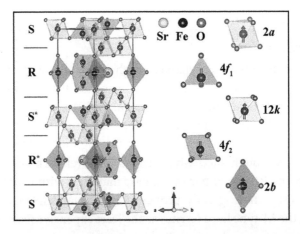

FIGURE 5.2 Typical crystal structure of the SrM ferrite and Fe^{3+} spin states at specific locations. Reprinted with permission from (Hu et al., 2020). Copyright (2020) Elsevier.

5.1.2 Crystal Structure and Magnetic Properties

M-type hexagonal $MFe_{12}O_{19}$ belongs to space group $P6_3/mmc$. As typically shown in Figure 5.2, the crystal structure of $SrFe_{12}O_{19}$ ferrite (SrM) is composed of four building blocks denoted as S, R, S* and R* (Hu et al., 2020). Block S is spinel with the formula of $[Fe^{3+}_6O_8]^{2+}$ and block R is hexagonal with the formula of $[M^{2+}Fe^{3+}_6O_{11}]^{2-}$. For S* and R*, the asterisk (*) designates a 180-degree rotation of spinel and hexagonal blocks circling the magnetically easy c-axis (Makovec et al., 2018). As observed from the Fe^{3+}

spin states at specific orbits, the magnetism of SrM ferrite arises from the Fe^{3+} distributed over five crystallographically nonequivalent sites in the unit cell, including one tetrahedral site ($4f_1$), one bipyramidal site (2b) and three octahedral sites (12k, $4f_2$ and 2a). Out of five nonequivalent sublattices, 2a, 2b and 12k sites are occupied by eight Fe^{3+} cations with upward spins, whereas $4f_1$ and $4f_2$ sites accommodate four Fe^{3+} cations with downward spins (Kohn et al., 1971). The super-exchange interaction between these parallel (2a, 12k, 2b) and antiparallel ($4f_1$, $4f_2$) sublattices via intermediate O^{2-} anions gives rise to ferrimagnetism (Güner et al., 2020; Lewis and Jimenez-Villacorta, 2013). Among different classes of M-type ferrites, larger divalent cations such as Ba^{2+} and Sr^{2+} are indispensable, which induces slight lattice perturbation and large magnetocrystalline anisotropy H_A. Table 5.1 summarizes the basic physical properties of the M-type ferrites, including density ρ, melting point T_m, lattice parameters a, c and V, saturation magnetization M_s, magnetocrystalline anisotropy constant K_1 and Curie temperature T_C (Pullar, 2012). Compared to the $BaFe_{12}O_{19}$ (BaM) with theoretical density of 5.295 g/cm³, SrM possesses a smaller density of 5.101 g/cm³. The radius of Pb^{2+} is in-between of those for Ba^{2+} and Sr^{2+}, while the Pb is heavier than Ba, generating a higher ρ of 5.708 g/cm³ for the $PbFe_{12}O_{19}$ (PbM). Compared with the PbM with drawbacks of relatively low magnetic performance and toxicity, the BaM and SrM with respective M_s of 380 and 370 kA/m are massively produced.

Figure 5.3 compares the thermal stability of different permanent magnets (Gutfleisch et al., 2011). Although the NdFeB and NdDyFeCoB possess superior room-temperature magnetic properties (particularly for the

TABLE 5.1 Basic Physical Characteristics of the M-Type Ferrites (Pullar, 2012)

Materials	$BaFe_{12}O_{19}$	$SrFe_{12}O_{19}$	$PbFe_{12}O_{19}$
ρ (g/cm³)	5.295	5.101	5.708
T_m (K)	1611	1692	1538
a (Å)	5.8876	5.8844	5.8941
c (Å)	23.1885	23.0632	23.0984
V (Å³)	696.2	691.6	694.9
M_s (kA/m, 293 K)	370	380	320
K_1 (kJ/m³, 293 K)	330	350	220
T_C (K)	725	732	718

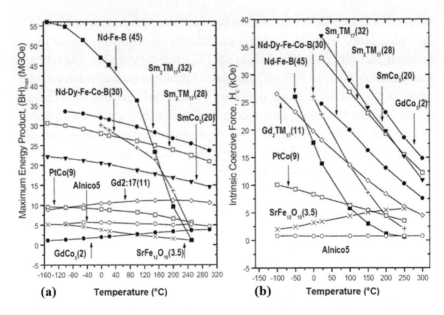

FIGURE 5.3 (a) Maximum energy product and (b) intrinsic coercivity as a function of temperature T for different permanent magnets. The numbers in the parentheses refers to the $(BH)_{max}$ at 25 °C. Reprinted with permission from (Gutfleisch et al., 2011). Copyright (2011) John Wiley and Sons.

giant maximum energy product shown in Figure 5.3a), severe thermal degradation occurs upon heating to 250 °C. By contrast, most of the hard ferrites exhibit nonlinear B–H curve at room temperature, while can be offset by the positive temperature coefficient of coercivity. Figure 5.3b shows that the coercivity of the $SrFe_{12}O_{19}$ ferrite can be enhanced from 3.2 kOe (0 °C) to 5.9 kOe (250 °C), which is advantageous for the motor applications. Except for ferrite, other hard magnetic materials disappointingly have negative coercivity coefficients upon higher service temperature.

5.1.3 Synthesis Techniques

Synthesis techniques of the M-type hard ferrites mainly include the standard ceramic method, sol-gel, coprecipitation and hydrothermal synthesis, which will be discussed in the following sections.

5.1.3.1 Standard Ceramic Method

Typical flow chart of the standard ceramic method is shown in Figure 5.4 (Huang et al., 2020). Firstly, the raw materials are weighed and mixed,

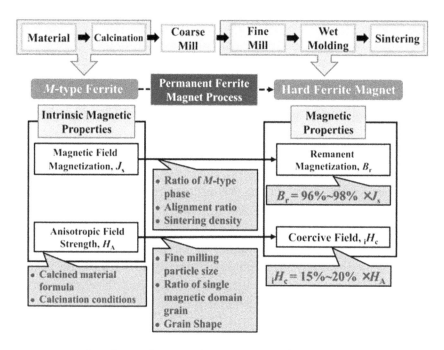

FIGURE 5.4 Correlations between the magnetic performance and related factors in the manufacturing process of the hard ferrites. Reprinted with permission from (Huang et al., 2020). Copyright (2020) Elsevier.

followed by pre-sintering calcination to produce the desired hexagonal phase. After that, the main composition and intrinsic properties including J_s and H_A of M-type ferrites are determined. Secondly, the calcined powders combined with secondary additives are coarsely and finely milled to guarantee finer and more homogeneous mixture. The coercivity of hard ferrite is greatly affected by the intensity and duration of milling. Optimum milling yields controllable particle size below the critical size of single domain, which is 0.5–1 µm for SrM and BaM ferrites. Thirdly, to take the advantage of the uniaxial anisotropy of hard ferrite, the wet molding is usually assisted by magnetic field to rotate the magnetic moment of crystal grains along the direction of applied field. Since the remanence B_r is roughly proportional to the alignment degree, the oriented ferrites usually possess larger B_r. Final sintering is critical for desired microstructure with grain size in the order of magnitude of ~1 µm. Hence optimum sintering should balance the complex interplay between dense microstructure and grain growth.

Note that the standard ceramic method usually requires high-temperature sintering for long duration and easily causes the coarse grains from a few microns to dozens of microns, which is harmful to the coercivity. Therefore grain growth inhibitor such as SiO_2, CaO and La_2O_3 have been introduced to produce fully-dense, well-oriented and fine-grained ferrite approaching the single-domain size (Hussain et al., 2006; Topfer et al., 2005; Onreabroy et al., 2012). Conventional ceramic sintering method has been widely applied in large-scale industrial production for more than half a century, despite the fact that the harmful impurities can be easily introduced during the milling process, and the high-temperature sintering is energy-consuming. Till now, the remanence approaches 96–98% of the theoretical J_s value, ascribed to the large fraction of M-type phase, high alignment degree and full densification (Figure 5.4). However, the achieved coercivity of hard ferrite is only 15–20% of the theoretical H_A limited by the particle size, the ratio of single-domain grains, and the grain shape.

5.1.3.2 Sol-Gel Method

The sol-gel procedures usually involve the reaction of metal salts under alkali conditions to form colloidal sol, which condenses into a gel followed by combustion to produce fine-grained M-type ferrites with particle size below 400 nm (Sankaranarayanan et al., 1993; Eikeland et al., 2018). Zhong et al. have systematically studied the formation mechanism, morphology and properties of the BaM ferrites via sol-gel, and found that the pre-heat treatment of the gel at 400–500 °C is critical to eliminate the α-Fe_2O_3 impurity and permit the low-temperature annealing at 900 °C (Zhong et al., 1997). As a result, the ultrafine $BaFe_{12}O_{19}$ single phase has been achieved with an ultimate magnetization of 70 emu/g and H_{cj} of 5.95 kOe. Recently, the SrM nanocrystalline powders with larger H_{cj} of 6.45 kOe have been prepared by mixing the $Fe(NO_3)_2 \cdot 9H_2O$ and $Sr(NO_3)_2$ precursors with stoichiometric ratio of 12:1, with $C_2H_4(OH)_2/C_6H_8O_7$ agents and ammonia to prepare the gel, followed by the citrate combustion reaction and thermal treatment (Guzmán-Mínguez et al., 2021). The high coercivity has been ascribed to the two-step post-synthesis processing, i.e. ball milling in wet ethanol and high-speed mixing to reduce the particle agglomeration.

5.1.3.3 Coprecipitation Method

Since the early 1960s, the chemical coprecipitation method has been used to produce the hexagonal ferrites with submicron particle size, which requires a non-stoichiometric Fe-deficient composition. The Fe:Ba mole ratio for BaM is 8:1–11:1, while the Fe:Sr mole ratio for SrM is lower than 9:1, rather than the stoichiometric value of 12:1 (Davoodi and Hashemi, 2012; Rashad et al., 2008). The aqueous solutions of raw materials are mixed according to a specific ratio, and then added with a precipitation agent such as alkali solution. The resultant precipitate is then centrifuged, segregated, washed and dried at a relatively low heating temperature. Such coprecipitation route is inexpensive and efficient, but with one shortcoming of α-Fe_2O_3 formation. To suppress the formation of secondary α-Fe_2O_3, the introduction of surfactants like cetyl-trimethyl ammonium bromide (CTAB) has been explored during the coprecipitation of SrM (Lu et al., 2011; Chen et al., 2012; Harikrishnan et al., 2016). The SrM platelets with high H_{cj} of 6.3 kOe have been fabricated by 1 wt% CTAB-assisted coprecipitation and 900 °C annealing (Harikrishnan et al., 2016). The single-phase BaM nanopowders have been produced with the addition of triton X-100 surfactant, giving rise to higher magnetic properties (H_{cj} = 4.58 kOe, M_s = 46.8 emu/g) than the un-doped one (H_{cj} = 3.92 kOe, M_s = 28.8 emu/g) (Rashad et al., 2008). Effects of various fabrication parameters have also been investigated to refine the microstructure and phase constitution of the ferrite. For instance, increased pH from 9 to 13 or decreased Fe:Sr ratio from 12:1 to 9:1 has been demonstrated to facilitate the formation of SrM nanoparticles with maximized M_s of 51 emu/g (Davoodi and Hashemi, 2012).

5.1.3.4 Hydrothermal Synthesis

The hydrothermal method also involves the use of metal salts under alkali conditions, which usually gives rise to mixed phases with unreacted precursors such as Sr-rich phase, and requires washing with dilute acetic acid. A modified hydrothermal route has been proposed to prepare the superparamagnetic and disc-shaped BaM nanoparticles to be ~10-nm wide and 3-nm thick (Drofenik et al., 2011; Primc et al., 2009). The formation temperature can be reduced by increasing the concentration of hydroxyl ions (OH^-) relative to the metal ions (Ba^{2+} and Fe^{3+}), in extreme case below 150 °C. Microwave irradiation has been combined with the hydrothermal method for faster heating rate (Bilecka and Niederberger, 2010), which yields a higher coercivity of 5.85 kOe for nanostructured SrM (Grindi et al., 2018).

5.1.3.5 Other Methods

Other chemical routes have also been developed to synthesize the M-type hard ferrites. Compared with the standard ceramic method requiring high calcination temperatures of 1200–1350 °C, the molten salt method can be conducted at lower reaction temperatures of 800–1100 °C for shorter time to impede particle agglomeration and abnormal crystal growth (He et al., 2014). Self-propagating high-temperature synthesis serves as a fast and inexpensive method for producing high-quality M-type ferrites even in air (Yang et al., 2009). Also, simplified auto-combustion provides a high-efficient and energy-saving alternative to prepare M-type nanoparticles (Han et al., 2019).

5.1.4 Substituted M-Type Ferrites

The magnetic properties of M-type hard ferrites are intrinsically linked to their composition and crystalline structure. Given the formula of M-type hexaferrite as $MFe_{12}O_{19}$, the strategies of ion substitution are mainly divided into two categories based on whether M^{2+} or Fe^{3+} is replaced.

Alkali metal ions such as Na^+, K^+, Rb^+, Ca^{2+} and REs are common substitutes for the M^{2+} ions (Seifert et al., 2009; Rehman Ur et al., 2019). Among them, substitution of the divalent M^{2+} with trivalent La^{3+} is commonly used, which is associated with valence change of one Fe^{3+} to Fe^{2+} per formula unit (Küpferling et al., 2005; Lotgering et al., 1980). The resultant La-hexaferrite exhibits relatively larger H_A at room temperature (Hessien et al., 2020). For the single-phase $Sr_{1-x}La_xFe_{12}O_{19}$ (x = 0–0.2) via the molten salt assisted sol-gel technique, the La^{3+} concentration significantly affects the grain growth and grain orientation (Lei et al., 2016). For the phase stability of Ce-substituted M-type ferrite, the oxygen pressure should be precisely controlled due to the fluctuation of the Ce valence (Waki et al., 2020).

Ions such as Al^{3+}, Zn^{2+}, Co^{2+}, Cr^{3+} and Ga^{3+} are common candidates for the substitution of Fe^{3+}. The accommodations of different ions in the dissimilar symmetry sites provide numerous possibilities for tailoring the magnetic properties of hard ferrites. Advanced techniques including Mössbauer spectroscopy, nuclear magnetic resonance spectroscopy and neutron diffraction have been employed to determine the preferential site occupation of different ions. It has been revealed that the Al^{3+} ions preferentially substitute the 2a sites, followed by the 12k sites. All the Fe^{3+} ions at these locations exhibit upward magnetic moments, and the substitution of nonmagnetic Al^{3+} reduces the net magnetic moment for decreased M_s. Similarly, Al^{3+} substitution also decreases the K_1. According to the relation of $H_A = 2K_1/M_s$, since M_s decreases more rapidly than the decrement of K_1, the H_A can then be enhanced with increasing Al^{3+} substitution (Rhein et al., 2017). For instance, the SrM ferrite exhibits gradually raised coercivity with increasing Al^{3+} substitution, which maximizes at 17.57 kOe for the $SrAl_4Fe_8O_{19}$ (Wang et al., 2012). Meanwhile, the replacement of Al^{3+} for Fe^{3+} weakens the super-exchange between Fe^{3+} ions, generating linear T_C decrement (Dahal et al., 2014). Doping of nonmagnetic Zn^{2+} has been found to enhance the M_s owing to the preferential $4f_1$ occupancy (Auwal et al., 2017), at the sacrifice of H_A and H_{cj}.

Beside single-ion replacement, La-Co co-substitution has been developed as a common strategy to prepare high-quality M-type ferrites, featuring coercivity increment without sacrificing the remanence and energy product (Liu et al., 2006; Liu et al., 2018). Represented by the commercial ferrites from Hengdian Group DMEGC Magnetics Co., Ltd, the high-grade DM4550 ferrite with La-Co co-substitution exhibits large B_r of 4.5 ± 0.1 kG, H_{cj} of 4.9 ± 0.2 kOe, and $(BH)_{max}$ of 4.7 ± 0.2 MGOe. Moreover, high-performance sub-micrometer SrLaCoCaM ferrite with $(BH)_{max}$ = 5.46 MGOe and H_{cj} = 5.6 kOe has been prepared (Li et al., 2016). Site occupation studies reveal that the Co^{2+} ions preferentially enter the 2a and 2b sites in the BaCoM hexaferrite (Mahadevan et al., 2020). Contrastly for the SrLaCoM hexaferrite, the Co^{2+} ions preferentially enter the $4f_1$ sites with the remaining Co^{2+} ions distributing between the 12k and 2a sites (Oura et al., 2018; Yoshinori et al., 2011; Nakamura et al., 2019; Sakai et al., 2018). Such site preferences generate positive contribution to the K_1 for increased H_A and H_{cj} while maintaining the B_r.

Other co-substitution systems have also been reported with high magnetic performance, including Mo-Zn (Mahmood et al., 2015), RE-Zn (Liu et al., 2019a; Almessiere et al., 2019), and RE-Al (Rai et al., 2013). Effects of RE-Al co-doping on the grain size and magnetic parameters of $Sr_{0.9}RE_{0.1}Fe_{10}Al_2O_{19}$ are shown in Table 5.2 (Rai et al., 2013). The M_s decreases mainly due to the valance change from Fe^{3+} to Fe^{2+} and the non-collinear spin arrangement of the magnetic moments. Similar decrement in the T_C is also observed due to the weakened super-exchange

TABLE 5.2 Grain Size and Magnetic Parameters of the RE-Al Co-Doped $Sr_{0.9}RE_{0.1}Fe_{10}Al_2O_{19}$ Ferrites Compared to the RE-Free and Al-Free Ones (Rai et al., 2013)

Samples	Grain Size (nm)	M_s (emu/g)	H_{cj} (kOe)	T_C (K)
$SrFe_{12}O_{19}$	91 ± 16	59.33	4.29	737
$SrFe_{10}Al_2O_{19}$	94 ± 17	36.50	7.40	629
$Sr_{0.9}Y_{0.1}Fe_{10}Al_2O_{19}$	95 ± 17	36.0	8.87	622
$Sr_{0.9}La_{0.1}Fe_{10}Al_2O_{19}$	120 ± 20	33.7	8.81	617
$Sr_{0.9}Ce_{0.1}Fe_{10}Al_2O_{19}$	101 ± 16	32.7	9.48	620
$Sr_{0.9}Pr_{0.1}Fe_{10}Al_2O_{19}$	99 ± 19	30.8	11.0	622
$Sr_{0.9}Nd_{0.1}Fe_{10}Al_2O_{19}$	106 ± 24	34.0	10.0	615
$Sr_{0.9}Sm_{0.1}Fe_{10}Al_2O_{19}$	138 ± 22	33.6	9.34	617
$Sr_{0.9}Gd_{0.1}Fe_{10}Al_2O_{19}$	148 ± 25	33.9	9.62	617

interactions. The H_{cj} of all the substituted samples are enhanced, among which the Pr^{3+} doping gives rise to the highest H_{cj} of 11.0 kOe.

5.1.5 Prospects and Future Challenges

The RE criticality offers great opportunities for the design of low-cost non-RE electrical machines based on the M-type ferrites, such as the permanent magnet brushless machines (PMBMs) (Jeong and Hur, 2017), PM-assisted synchronous reluctance machines (PMa-SynRMs) (Wu et al., 2017), PM flux-switching machines (PMFSMs) (Zhao et al., 2016), and PM vernier machines (PMVMs) (Chen et al., 2018). The combination of ferrites and other materials also attracts fundamental and application interests for batteries, spintronics, and microwave communications (Mann, 2011). Recently, by introducing BaM nanoparticles, a composite hard magnetic ink is successfully prepared, which can be screen-printed with maximum operating temperature up to 300 °C, and capable of medical and stretchable electronic application (Smith et al., 2020). $BaFe_{12}O_{19}$ doping agent is also an effective way to solve the nanoparticle agglomeration in nanocomposites (Liu et al., 2019b).

For over 70 years, the quality of commercial ferrites is enhanced through accumulated fundamental and technological knowledge. The experimental $(BH)_{max}$ is close to the theoretical value, while the coercivity require further improvements. The optimum preparation methods and the complex mechanism of multi-ions substitution should remain an area of intensive research interest.

5.2 ALNICO

5.2.1 Introduction

The name of Alnico follows its main compositions including Fe, Al, Ni and Co, with minor additives such as Cu, Ti, Nb and Si. The magnetic strength of Alnico stems from the spinodal decomposition of homogeneous phase α into bcc ferromagnetic FeCo-rich α_1 and bcc nonmagnetic AlNi-rich α_2 during heat treatment (Figure 5.5a) (Heidenreich and Nesbitt, 1952; Coey, 2010). Figures 5.5b and c illustrate the bcc crystal structures of the FeCo-rich and AlNi-rich phases (Nesbitt and Heidenreich, 1952). The M_s and T_C mainly derive from the FeCo-rich phase, while the H_A is insufficient to induce suitable coercivity. Instead, the coercivity of Alnico magnet can be compensated by the shape anisotropy of nanosized α_1 rods finely dispersed in the nonmagnetic α_2 matrix (Nesbitt and Williams, 1955; Cahn, 1963).

5.2.2 Timeline of Alnico Development

The timeline of Alnico development should be traced back to the discovery of FeNiAl system in the early 1930s with B_r of 9.4 kG and H_{cj} of 0.43 kOe as the best available magnetic steels at that time (Mishima, 1931; 1932). The following decades witnessed improved $(BH)_{max}$ by optimizing the composition design and thermal treatment based on empirical studies. The addition of Co induces the FeCo-rich precipitates, resulting in enhanced $(BH)_{max}$ from 0.38 to 1.90 MGOe (Snoek, 1939). The introduction of Cu accelerates the spinodal decomposition, while minor Ti, Nb and Si additives effectively inhibit the columnar grain growth by forming oxides in the melt for enhanced magnetic properties. The early Alnico 1,

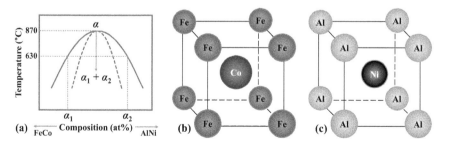

FIGURE 5.5 (a) Spinodal decomposition schematic of Alnico, as well as the crystal structures of the (b) FeCo-rich and (c) AlNi-rich phases.

2 and 3 magnets with low coercivity of 0.6–1.3 kOe consists of randomly oriented FeCo parallelepipeds. Controlled directional solidification from a temperature above 1100 °C and heat treatment at 550–700 °C for several hours improve the magnetic properties. The thermal magnetic treatment gives rise to oriented and elongated FeCo precipitates with the <001> texture, producing higher shape anisotropy (Oliver and Shedden, 1938). Since then, the anisotropic Alnico exhibiting large coercivity and energy product became available. A series grades of commercial Alnico have already been developed in the 1960s, which can be produced from over 100 kg by casting and down to a few grams by sintering (Kirchmayr, 1996), once dominating the permanent magnet market during mid-20th century. From the 1970s with the discovery of powerful SmCo and NdFeB permanent magnets with higher magnetocrystalline anisotropy, the market share of Alnico started to decrease. However, attractive technical magnetic characteristics such as large magnetization (7.0–13.5 kG), high T_C (over 1073 K), and near-zero temperature coefficients (α of −0.02%/K, β of −0.015%/K) contribute to the irreplaceable applications of the Alnico in various meters and aerospace industry.

Table 5.3 shows the current manufacturing standards of compositions and magnetic properties for representative Alnico alloys (Zhou et al., 2014; Cui et al., 2018; Mohapatra and Liu, 2018). The average compositions of α_1/α_2 phases determined using atom probe have been summarized in Table 5.4 (Zhou et al., 2014; Cui et al., 2018). Cast Alnico 1–3 are magnetically isotropic with randomly oriented grains. Cast Alnico 5 and Alnico 5–7 possess the highest Fe content, contributing to the higher B_r of 12.8 and 13.5 kG but also the lower H_{cj} of 0.64 and 0.74 kOe, respectively. Higher $(BH)_{max}$ of cast Alnico 5, 5–7, 6, 8, 8HC and 9 is due to their anisotropic texture along <100> axis by high-temperature-gradient solidification. Particularly, Alnico 9 after the thermal magnetic annealing was developed with high H_{cj} of 1.5 kOe, as well as the highest $(BH)_{max}$ of 10.5 MGOe. Sintered Alnico 2, 5, 6 and 8 have also been manufactured with moderate decrement in magnetic properties compared to the cast counterparts.

In the last decade, advances in computational modeling, nanoscale and atomic scale characterization instruments have made it possible to reinvestigate the magnetic behavior of Alnico. Theoretical $(BH)_{max}$ of 21.4 MGOe has been estimated for Alnico 9 magnet with optimal microstructure (Zhou et al., 2014). Figure 5.6 compares the theoretical, experimental and production values of $(BH)_{max}$ for several well-known permanent

Rare-Earth-Free Hard Magnetic Materials ▪ 73

TABLE 5.3 Summary of Compositions and Magnetic Properties for Commercial Alnico (Mohapatra and Liu, 2018)

Grade	Co (at%)	Al (at%)	Ni (at%)	Ti (at%)	$(BH)_{max}$ (MGOe)	B_r (kG)	H_{cj} (kOe)	Operating Temperature (°C)	Texture
Cast Alnico 1	5	22.3	18.8	0	1.4	7.2	0.48	450	Isotropic
Cast Alnico 2	12	18.6	17.5	0	1.7	7.5	0.58	450	
Cast Alnico 3	0	22.3	21.4	0	1.35	7	0.5	450	
Cast Alnico 5	21.4	15.6	12.5	0	5.5	12.8	0.64	525	Anisotropic
Cast Alnico 5-7	21.4	15.6	12.5	0	7.5	13.5	0.74	525	
Cast Alnico 6	21.4	15.6	14.3	1.1	3.9	10.5	0.8	525	
Cast Alnico 8	31.5	13.8	13.6	5.5	5.3	8.2	1.86	550	
Cast Alnico 8HC	33.7	15.5	12.5	8.7	5	7.2	2.17	550	
Cast Alnico 9	31.5	13.8	13.6	5.5	10.5	11.2	1.5	550	
Sintered Alnico 2	11.3	19	16.6	0	1.5	7.1	0.57	450	
Sintered Alnico 5	21.4	15.6	12.5	0	3.9	10.9	0.63	550	
Sintered Alnico 6	21.4	15.6	13.4	1.1	2.9	9.4	0.82	525	
Sintered Alnico 8	31.5	13.8	13.6	5.5	4	7.4	1.69	550	

TABLE 5.4 Compositions of FeCo-Rich α_1 and d AlNi-Rich α_2 Phases for Alnico 5-7, 8 and 9 (Zhou et al., 2014)

Composition	α_1 Phase (at%)			α_2 Phase (at%)		
	Alnico 5-7	Alnico 8	Alnico 9	Alnico 5-7	Alnico 8	Alnico 9
Fe	66.7	46.4	50.9	13.9	9.1	12.3
Co	24.5	40.5	35.8	16.4	28.2	26.0
Ni	3.1	7.7	5.3	30.5	24.1	23.1
Al	4.9	4.7	5.6	34.5	23.2	25.0
Cu	0.3	0.3	0.4	4.3	1.6	1.6
Ti	0	0.5	0.9	0	13.6	10.7

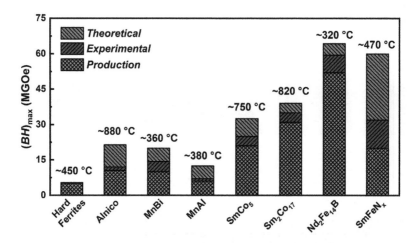

FIGURE 5.6 Comparisons of the theoretical, experimental and commercial production values of $(BH)_{max}$ for typical permanent magnets.

magnets with different T_C. Among them, the Alnico appears to be one of the most promising permanent magnets for high-temperature applications. It is also worth noting that the experimental and production $(BH)_{max}$ of ~12.2 MGOe and ~10.5 MGOe are only nearly half of the theoretical value for the Alnico. A target $(BH)_{max}$ of ~20 MGOe at 180 °C is a performance set that allows Alnico to compete with RE-based permanent magnets for electric motor applications at elevated temperatures.

Figure 5.7 shows the advanced high-angle-annular-dark-field scanning transmission electron microscopic (HAADF-STEM) images of the commercial Alnico 5-7, 8 and 9. Compared to the brick-and-mortar-like structure

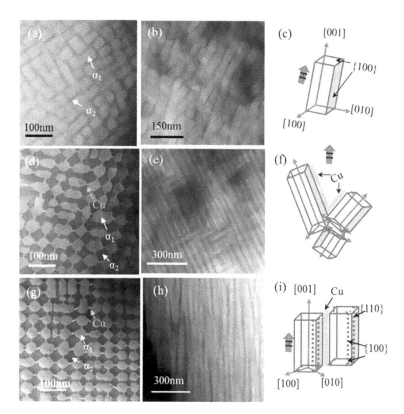

FIGURE 5.7 HAADF-STEM images and schematic of α_1 phase morphology of different Alnico alloys, including (a) Alnico 5-7 (transverse), (b) Alnico 5-7 (longitudinal), (d) Alnico 8 (transverse), (e) Alnico 8 (longitudinal), (g) Alnico 9 (transverse), (h) Alnico 9 (longitudinal). Models of the α_1 phase in Alnico 5-7, 8 and 9 are shown in (c), (f) and (i). Reprinted with permission from (Zhou et al., 2014). Copyright (2014) Elsevier.

with uniform Cu distribution in the α_2 phase for the Alnico 5-7, the Alnico 8 and 9 with higher coercivity exhibit a mosaic-like structure with Cu segregation between the α_1/α_2 boundaries (Zhou et al., 2014). During thermal magnetic treatment, the FeCo-rich phase is initially coarsened to a certain size, followed by splitting into smaller ones, resulting in more homogeneous and finer microstructure (Chu et al., 2000; 2002). Furthermore, phase field simulation reveals that the larger external magnetic field energy E_{ext} generates smaller spinodal decomposed magnetic phases (Sun et al., 2013). Besides the initial thermal magnetic treatment, subsequent low-temperature and long-time heat treatment also plays a crucial role for enhanced coercivity, which is ascribed to the modified α_1/α_2 interface by the elongated and coalesced Cu-enriched clusters (Zhou et al., 2017).

5.2.3 Prospects and Future Challenges

Till now there are still several drawbacks for the Alnico, including the low coercivity limited by the shape anisotropy mechanism, the high cost almost equivalent to the RE-based magnets, as well as the prolonged high temperature annealing and fine machining procedures. To alleviate the high material cost caused by the expensive Co, the so-called cobalt-lean Alnico has been predicted by Rigid Band Approximation method (Palasyuk et al., 2017), which is supposed to possess comparable magnetic properties to traditional Co-rich Alnico. To shorten the thermal treatment process and avoid additional material waste, net-shape additive manufacturing (AM) of Alnico has been developed with negligible impact on the magnetic performance (White et al., 2017; Rottmann et al., 2021). The remanence of the AM processed Alnico reaches 9 kG, which is close to the value of the anisotropic cast Alnico 9 with the same composition. It provides new perspectives toward high cost-performance Alnico.

Progresses on modern characterization and simulation also assist to determine the mechanism of spinodal decomposition, providing more comprehensive understanding on the composition-microstructure-property correlation. Ideal Alnico closer to the theoretical $(BH)_{max}$ of ~20 MGOe may be attainable, if potential pathways of the anisotropic magnet are provided, involving (i) strictly aligned grains in the initial alloy under applied magnetic field; (ii) optimal thermal magnetic treatment for fine spinodal structure; and (iii) appropriate subsequent low-temperature treatment and composition design for coherent α_1/α_2 interface.

To sum up, the currently available hard ferrites and Alnico described in this chapter still maintain competitive advantages for commercial applications. From the economic and environmental perspectives, the development of rare-earth-free hard magnetic materials will remain an important pursuit for the magnetic materials designers and engineers, with the consideration of delicate compositional and structural design, as well as novel processing methods.

REFERENCES

Almessiere, M. A., Y. Slimani, and A. Baykal. 2019. Impact of Nd-Zn co-substitution on microstructure and magnetic properties of $SrFe_{12}O_{19}$ nanohexaferrite. *Ceramics International* 45(1):963–9.

Auwal, I. A., A. D. Korkmaz, M. D. Amir, S. M. Asiri, A. Baykal, H. Güngüneş, and S. E. Shirstah. 2017. Mössbauer analysis and cation distribution of Zn substituted $BaFe_{12}O_{19}$ hexaferrites. *Journal of Superconductivity and Novel Magnetism* 31(1):151–6.

Bilecka, I., and M. Niederberger. 2010. Microwave chemistry for inorganic nanomaterials synthesis. *Nanoscale* 2(8):1358–74.

Cahn, J. W. 1963. Magnetic aging of spinodal alloys. *Journal of Applied Physics* 34(12):3581–6.

Chen, D. Y., Y. Y. Meng, D. C. Zeng, Z. W. Liu, H. Y. Yu, and X. C. Zhong. 2012. CTAB-assisted low-temperature synthesis of $SrFe_{12}O_{19}$ ultrathin hexagonal platelets and its formation mechanism. *Materials Letters* 76:84–6.

Chen, Y. Y., Y. Ding, X. J. Li, and X. H. Zhu. 2018. Design and analysis of less-rare-earth double-stator modulated machine considering multioperation conditions. *IEEE Transactions on Applied Superconductivity* 28(3):5201405.

Chu, W. G., W. D. Fei, X. H. Li, and D. Z. Yang, 2002. A study on the microstructure of Alnico8 alloy thermomagnetically treated at various temperatures. *Materials Chemistry and Physics* 73(2–3):290–4.

Chu, W. G., W. D. Fei, X. H. Li, D. Z. Yang, and J. L. Wang, 2000. Evolution of Fe-Co rich particles in Alnico 8 alloy thermomagnetically treated at 800 °C. *Materials Science and Technology* 16(9):1023–8.

Coey, J. M. 2010. *Magnetism and magnetic materials*. New York: Cambridge University Press.

Cui, J., M. Kramer, L. Zhou, F. Liu, A. Gabay, G. Hadjipanayis, B. Balasubramanian, and D. Sellmyer, 2018. Current progress and future challenges in rare-earth-free permanent magnets. *Acta Materialia* 158:118–37.

Dahal, J. N., L. Wang, S. R. Mishra, V. V. Nguyen, and J. P. Liu. 2014. Synthesis and magnetic properties of $SrFe_{12-x-y}Al_xCo_yO_{19}$ nanocomposites prepared via autocombustion technique. *Journal of Alloys and Compounds* 595:213–20.

Davoodi, A., and B. Hashemi. 2012. Investigation of the effective parameters on the synthesis of strontium hexaferrite nanoparticles by chemical coprecipitation method. *Journal of Alloys and Compounds* 512(1):179–84.

Drofenik, M., I. Ban, D. Makovec, A. Žnidaršič, Z. Jagličić, D. Hanžel, and D. Lisjak. 2011. The hydrothermal synthesis of super-paramagnetic barium hexaferrite particles. *Materials Chemistry and Physics* 127(3):415–9.

Eikeland, A. Z., M. Stingaciu, A. H. Mamakhel, M. Saura-Muzquiz, and M. Christensen. 2018. Enhancement of magnetic properties through morphology control of $SrFe_{12}O_{19}$ nanocrystallites. *Scientific Reports* 8(1):7325.

Grindi, B., Z. Beji, G. Viau, and A. BenAli. 2018. Microwave-assisted synthesis and magnetic properties of M-$SrFe_{12}O_{19}$ nanoparticles. *Journal of Magnetism and Magnetic Materials* 449:119–26.

Güner, S., M. A. Almessiere, Y. Slimani, A. Baykal, and I. Ercan. 2020. Microstructure, magnetic and optical properties of Nb^{3+} and Y^{3+} ions co-substituted Sr hexaferrites. *Ceramics International* 46(4):4610–8.

Gutfleisch, O., M. A. Willard, E. Brück, C. H. Chen, S. G. Sankar, and J. P. Liu. 2011. Magnetic materials and devices for the 21st Century: stronger, lighter, and more energy efficient. *Advanced Materials* 23(7):821–42.

Guzmán-Mínguez, J. C., L. Moreno-Arche, C. Granados-Miralles, J. López-Sánchez, P. Marín, J. F. Fernández, and A. Quesada. 2021. Boosting the coercivity of $SrFe_{12}O_{19}$ nanocrystalline powders obtained using the citrate combustion synthesis method. *Journal of Physics D: Applied Physics* 54(1):014002.

Han, G. H., Y. Q. Liu, W. W. Yang, S. Geng, W. B. Cui, and Y. S. Yu. 2019. Fabrication, characterization, and magnetic properties of exchange-coupled porous $BaFe_8Al_4O_{19}/Co_{0.6}Zn_{0.4}Fe_2O_4$ nanocomposite magnets. *Nanoscale* 11(22):10629–35.

Harikrishnan, V., P. Saravanan, R. Ezhil Vizhi, D. Rajan Babu, V. T. P. Vinod, P. Kejzlar, and M. Černík. 2016. Effect of annealing temperature on the structural and magnetic properties of CTAB-capped $SrFe_{12}O_{19}$ platelets. *Journal of Magnetism and Magnetic Materials* 401:775–83.

He, X. M., W. Zhong, S. M. Yan, C. T. Au, L. Y. Lü, and Y. W. Du. 2014. The structure, morphology and magnetic properties of Sr-ferrite powder prepared by the molten-salt method. *Journal of Physics D: Applied Physics* 47(23):235002.

Heidenreich, R. D., and E. A. Nesbitt, 1952. Physical structure and magnetic anisotropy of alnico 5. Part I. *Journal of Applied Physics* 23(3):352–65.

Hessien, M. M., N. El-Bagoury, M. H. H. Mahmoud, M. Alsawat, A. K. Alanazi, and M. M. Rashad. 2020. Implementation of La^{3+} ion substituted M-type strontium hexaferrite powders for enhancement of magnetic properties. *Journal of Magnetism and Magnetic Materials* 498:166187.

Hussain, S., M. Anis-ur-Rehman, A. Maqsooda, and M. S. Awan. 2006. The effect of SiO_2 addition on structural, magnetic and electrical properties of strontium hexa-ferrites. *Journal of Crystal Growth* 297:403–10.

Hu, J. Y., C. C. Liu, X. C. Kan, X. S. Liu, S. J. Feng, Q. R. Lv, Y. J. Yang, W. Wang, M. Shezad, and K. M. Ur Rehman. 2020. Structure and magnetic performance of Gd substituted Sr-based hexaferrites. *Journal of Alloys and Compounds* 820:153180.

Huang, C. C., C. C. Mo, T. H. Hsiao, G. M. Chen, S. H. Lu, Y. H. Tai, H. H. Hsu, and C. H. Cheng. 2020. Preparation and magnetic properties of high performance Ca-Sr based M-type hexagonal ferrites. *Results in Materials* 8:100150.

Jeong, C. L., and J. Hur. 2017. Optimization design of PMSM with hybrid-type permanent magnet considering irreversible demagnetization. *IEEE Transactions on Magnetics* 53(11):8110904.

Kirchmayr, H. R. A. 1996. Permanent magnets and hard magnetic materials. *Journal of Physics D: Applied Physics* 29:2763.

Kohn, J. A., D. W. Eckart, and C. F. Cook. 1971. Crystallography of the hexagonal ferrites. *Science* 172:3982.

Küpferling, M., P. Novák, K. Knížek, M. W. Pieper, R. Grössinger, G. Wiesinger, and M. Reissner. 2005. Magnetism in La substituted Sr hexaferrite. *Journal of Applied Physics* 97(10):10F309.

Lei, C. L., S. L. Tang, and Y. W. Du. 2016. Synthesis of aligned La^{3+}-substituted Sr-ferrites via molten salt assisted sintering and their magnetic properties. *Ceramics International* 42(14):15511–6.
Lewis, L. H., and F. Jimenez-Villacorta, 2013. Perspectives on permanent magnetic materials for energy conversion and power generation. *Metallurgical and Materials Transactions A* 44:2–20.
Li, Y. P., Y. F. Wu, and D. X. Bao. 2016. Enhanced coercivity of La-Co substituted Sr-Ca hexaferrite fabricated by improved ceramics process. *Journal of Materials Science: Materials in Electronics* 27(5):4433–6.
Liu, C. C., X. C. Kan, F. Hu, X. S. Liu, S. J. Feng, J. Y. Hu, W. Wang, K. M. Ur Rehman, M. Shezad, C. Zhang, H. H. Li, S. Q. Zhou, and Q. Y. Wu. 2019a. Investigations of Ce-Zn co-substitution on crystal structure and ferrimagnetic properties of M-type strontium hexaferrites $Sr_{1-x}CeFe_{12-x}Zn_xO_{19}$ compounds. *Journal of Alloys and Compounds* 785:452–9.
Liu, C. C., X. S. Liu, S. J. Feng, K. M. Ur Rehman, M. L. Li, C. Zhang, H. H. Li, and X. Y. Meng. 2018. Effect of Y-La-Co substitution on microstructure and magnetic properties of M-type strontium hexagonal ferrites prepared by ceramic method. *Journal of Magnetism and Magnetic Materials* 445:1–5.
Liu, X. S., P. Hernández-Gómez, K. Huang, S. Q. Zhou, Y. Wang, X. Cai, H. J. Sun, and B. Ma. 2006. Research on La^{3+}-Co^{2+}-substituted strontium ferrite magnets for high intrinsic coercive force. *Journal of Magnetism and Magnetic Materials* 305(2):524–8.
Liu, Z. Y., J. L. Zhu, P. Wei, W. T. Zhu, W. Y. Zhao, A. L. Xia, D. Xu, Y. Lei, and J. Yu. 2019b. Candidate for magnetic doping agent and high-temperature thermoelectric performance enhancer: hard magnetic M-type $BaFe_{12}O_{19}$ nanometer suspension. *ACS Applied Materials & Interfaces* 11(49):45875–84.
Lotgering, F. K., P. R. Locher, and R. P. Stapele. 1980. Anisotropy of hexagonal ferrites with M, W and Y structures containing Fe^{3+} and Fe^{2+} as magnetic ions. *Journal of Physics and Chemistry of Solids* 41(5):481–7.
Lu, H. F., R. Y. Hong, and H. Z. Li. 2011. Influence of surfactants on co-precipitation synthesis of strontium ferrite. *Journal of Alloys and Compounds* 509:10127–31.
Mahadevan, S., V. Sathe, V. R. Reddy, and P. Sharma. 2020. Site occupation and magnetic studies in La-Co-substituted barium hexaferrite. *IEEE Transactions on Magnetics* 56(10):1–6.
Mahmood, S. H., A. N. Aloqaily, Y. Maswadeh, A. Awadallah, I. Bsoul, M. Awawdeh, and H. Juwhari. 2015. Effects of heat treatment on the phase evolution, structural, and magnetic properties of Mo-Zn doped M-type hexaferrites. *Solid State Phenomena* 232:65–92.
Mann, A. 2011. High-temperature superconductivity at 25: still in suspense. *Nature* 475:280–2.
Makovec, D., B. Belec, T. Gorsak, D. Lisjak, M. Komelj, G. DraZic, and S. Gyergyek. 2018. Discrete evolution of the crystal structure during the growth of Ba-hexaferrite nanoplatelets. *Nanoscale* 10(30):14480–91.

McCallum, R. W., L. Lewis, R. Skomski, M. J. Kramer, and I. E. Anderson. 2014. Practical aspects of modern and future permanent magnets. *Annual Review of Materials Research* 44(1):451–77.
Mohapatra, J., and J. P. Liu. 2018. Rare-earth-free permanent magnets: the past and future. In *Handbook of magnetic materials*, ed. E. Brück, 27:1–57. Amsterdam: North-Holland Publishing.
Mishima, T. 1931. Nickel-aluminum steel for permanent magnets. *Stahl Und Eisen* 53:79.
Mishima, T. 1932. Nickel-aluminum steel for permanent magnets. *Ohm* 19:353.
Nakamura, H., T. Waki, Y. Tabata, and C. Mény. 2019. Co site preference and site-selective substitution in La-Co co-substituted magnetoplumbite-type strontium ferrites probed by ^{59}Co nuclear magnetic resonance. *Journal of Physics: Materials* 2(1):015007.
Nesbitt, E. A., and H. J. Williams 1955. Shape and crystal anisotropy of Alnico 5. *Journal of Applied Physics* 26:1217.
Nesbitt, E. A., and R. D. Heidenreich, 1952. Physical structure and magnetic anisotropy of Alnico 5. Part II. *Journal of Applied Physics* 23:366.
Oliver D. A., and J. W. Shedden, 1938. Cooling of permanent magnet alloys in a constant magnetic field. *Nature* 142:209.
Onreabroy, W., K. Papato, G. Rujijanagul, K. Pengpat, and T. Tunkasiri. 2012. Study of strontium ferrites substituted by lanthanum on the structural and magnetic properties. *Ceramics International* 38S:S415–9.
Oura, M., N. Nagasawa, S. Ikeda, A. Shimoda, T. Waki, Y. Tabata, H. Nakamura, N. Hiraoka, and H. Kobayashi. 2018. ^{57}Fe Mössbauer and Co K_β x-ray emission spectroscopic investigations of La-Co and La substituted strontium hexaferrite. *Journal of Applied Physics* 123(3):033907.
Palasyuk, A., R. W. McCallum, I. E. Anderson, M. Kramer, L. Zhou, and W. Tang. 2017. Cobalt-lean alnico alloy. *U.S. Patent Application* 754:15/330.
Primc, D., D. Makovec, D. Lisjak, and M. Drofenik. 2009. Hydrothermal synthesis of ultrafine barium hexaferrite nanoparticles and the preparation of their stable suspensions. *Nanotechnology* 20(31):315605.
Rai, B. K., S. R. Mishra, V. V. Nguyen, and J. P. Liu. 2013. Synthesis and characterization of high coercivity rare-earth ion doped $Sr_{0.9}RE_{0.1}Fe_{10}Al_2O_{19}$ (RE: Y, La, Ce, Pr, Nd, Sm, and Gd). *Journal of Alloys and Compounds* 550:198–203.
Rashad, M. M., M. Radwan, and M. M. Hessien. 2008. Effect of Fe/Ba mole ratios and surface-active agents on the formation and magnetic properties of co-precipitated barium hexaferrite. *Journal of Alloys and Compounds* 453(1–2):304–8.
Rehman Ur, K. M., M. Riaz, X. S. Liu, M. W. Khan, Y. J. Yang, K. M. Batoo, S. F. Adil, and M. Khan. 2019. Magnetic properties of Ce doped M-type strontium hexaferrites synthesized by ceramic route. *Journal of Magnetism and Magnetic Materials* 474:83–9.
Rhein, F., R. Karmazin, M. Krispin, T. Reimann, and O. Gutfleisch. 2017. Enhancement of coercivity and saturation magnetization of Al^{3+} substituted M-type Sr-hexaferrites. *Journal of Alloys and Compounds* 690:979–85.

Rottmann, P. F., A. T. Polonsky, T. Francis, M. G. Emigh, M. Krispin, G. Rieger, M. P. Echlin, C. G. Levi, T. M. Pollock. 2021. TriBeam tomography and microstructure evolution in additively manufactured Alnico magnets. *Materials Today* 49:23–34.

Sakai, H., T. Hattori, Y. Tokunaga, S. Kambe, H. Ueda, Y. Tanioku, C. Michioka, K. Yoshimura, K. Takao, A. Shimoda, T. Waki, Y. Tabata, and H. Nakamura. 2018. Occupation sites and valence states of Co dopants in (La, Co)-codoped M-type Sr ferrite: ^{57}Fe and ^{59}Co nuclear magnetic resonance studies. *Physical Review B* 98(6):064403.

Sankaranarayanan, V. K., Q. A. Pankhurst, D. P. E. Dickson, and C. E. Johnson. 1993. Ultrafine particles of barium ferrite from a citrate precursor. *Journal of Magnetism and Magnetic Materials* 120:73–5.

Seifert, D., J. Töpfer, F. Langenhorst, J. M. Le Breton, H. Chiron, and L. Lechevallier. 2009. Synthesis and magnetic properties of La-substituted M-type Sr hexaferrites. *Journal of Magnetism and Magnetic Materials* 321(24):4045–51.

Smith, C. S., K. Sondhi, S. C. Mills, J. S. Andrew, Z. H. Fan, T. Nishida, and D. P. Arnold. 2020. Screen-printable and stretchable hard magnetic ink formulated from barium hexaferrite nanoparticles. *Journal of Materials Chemistry C* 8(35):12133–9.

Snoek, J. L. 1939. Magnetic studies in the ternary system FeNiAl. *Physica* 6:321–31.

Sugimoto, M. 1999. The past, present, and future of ferrites. *Journal of the American Ceramic Society* 82(2):269–80.

Sun, X. Y., C. L. Chen, L. Yang, L. X. Lv, S. Atroshenko, W. Z. Shao, X. D. Sun, and L. Zhen, 2013. Experimental study on modulated structure in Alnico alloys under high magnetic field and comparison with phase-field simulation. *Journal of Magnetism and Magnetic Materials* 348:27–32.

Topfer, J., S. Schwarzerb, S. Senzc, and D. Hesse. 2005. Influence of SiO$_2$ and CaO additions on the microstructure and magnetic properties of sintered Sr-hexaferrite. *Journal of the European Ceramic Society* 25:1681–8.

Waki, T., G. Inoue, Y. Tabata, and H. Nakamura. 2020. Phase stability of Ce-substituted magnetoplumbite-type Sr ferrite. *IEEE Transactions on Magnetics* 56(3):1–4.

Wang, H. Z., B. Yao, Y. Xu, Q. He, G. H. Wen, S. W. Long, J. Fan, G. D. Li, L. Shan, B. Liu, L. N. Jiang, and L. L. Gao. 2012. Improvement of the coercivity of strontium hexaferrite induced by substitution of Al^{3+} ions for Fe^{3+} ions. *Journal of Alloys and Compounds* 537:43–9.

Went, J. J., G. W. Rathenau, E. W. Gorter, and G. W. Van Oosterhout. 1952. Ferroxdure, a class of new permanent magnet materials. *Philips Technical Review* 13:194–208.

White, E. M. H., A. G. Kassen, E. Simsek, W. Tang, R. T. Ott, and I. E. Anderson. 2017. Net shape processing of alnico magnets by additive manufacturing. *IEEE Transaction on Magnetics* 53(11):2101606.

Wu, W. Y., X. Y. Zhu, L. Quan, D. Y. Fan, and Z. X. Xiang. 2017. Characteristic analysis of a less-rare-earth hybrid PM-assisted synchronous reluctance motor for EVs application. *AIP Advances* 7(5):056648.

Yang, X. F., Q. L. Li, J. X. Zhao, B. D. Li, and Y. F. Wang. 2009. Preparation and magnetic properties of controllable-morphologies nano-$SrFe_{12}O_{19}$ particles prepared by sol-gel self-propagation synthesis. *Journal of Alloys and Compounds* 475(1-2):312-5.

Yoshinori K., O. Etsushi, N. Takeshi, and N. Takashi. 2011. Cation distribution analysis of Sr-La-Co M-type ferrites by neutron diffraction, extended X-ray absorption fine structure and X-ray magnetic circular dichroism. *Journal of the Ceramic Society of Japan* 119(1388):285-90.

Zhao, W. X., J. Q. Zheng, J. B. Wang, G. H. Liu, J. X. Zhao, and Z. Y. Fang. 2016. Design and analysis of a linear permanent- magnet vernier machine with improved force density. *IEEE Transactions on Industrial Electronics* 63(4):2072-82.

Zhong, W., W. P. Ding, N. Zhang, J. M. Hong, Q. J. Yan, and Y. W. Du. 1997. Key step in synthesis of ultrafine $BaFe_{12}O_{19}$ by sol-gel technique. *Journal of Magnetism and Magnetic Materials* 168(1-2):196-202.

Zhou, L., M. K. Miller, P. Lu, L. Ke, R. Skomski, H. Dillon, Q. Xing, A. Palasyuk, M. McCartney, and D. Smith. 2014. Architecture and magnetism of alnico. *Acta Materialia* 74:224-33.

Zhou, L., W. Tang, L. Ke, W. Guo, J. D. Poplawsky, I. E. Anderson, and M. J. Kramer. 2017. Microstructural and magnetic property evolution with different heat treatment conditions in an alnico alloy. *Acta Materialia* 133:73-80.

FURTHER READING

Pullar, R. C. 2012. Hexagonal ferrites: A review of the synthesis, properties and applications of hexaferrite ceramics. *Progress in Materials Science* 57(7):1191-334.

Zhou, L., W. Tang, L. Ke, W. Guo, J. D. Poplawsky, I. E. Anderson, and M. J. Kramer. 2017. Microstructural and magnetic property evolution with different heat treatment conditions in an alnico alloy. *Acta Materialia* 133:73-80.

CHAPTER 6

Rare-Earth-Based Hard Magnetic Materials: SmCo

6.1 DEVELOPMENTS OF RECO HARD MAGNETIC MATERIALS

The discovery of SmCo magnets opens a new period of intensive and fruitful investigations on the 3d-4f transitional metallic compounds. In 1966, the YCo$_5$ intermetallic phase with uniaxial symmetry was found to exhibit an extremely high H_A of ~140 kOe (Hoffer and Strnat, 1966). Although the early experimental results of YCo$_5$ magnets with B_r = 3.68 kG, H_{cj} = 1.2 kOe and $(BH)_{max}$ of 1.1 MGOe were not encouraging, they stimulated the quest of other RECo intermetallic phases with strong 3d–4f coupling and large H_A (Buschow, 1988). In 1968, all the CaZn$_5$-type RECo intermetallic compounds were studied, leading to the advent of the first-generation SmCo$_5$ permanent magnets (Velge and Bushow, 1968). In 1970, SmCo$_5$ bulk magnets were fabricated using the liquid-phase-sintering technique, which became a standard process for SmCo and NdFeB magnets (Benz and Martin, 1970). Afterward, SmCo$_5$ sintered magnets exhibiting high $(BH)_{max}$ of 24–25 MGOe were prepared (Benz and Martin, 1972), as the typical value of current commercial SmCo$_5$. Sm$_2$Co$_{17}$ intermetallic phase with higher Co concentration and thus higher $(BH)_{max}$ over the earlier SmCo$_5$ was the next important step forward, which soon evolved as the second generation of RE-based permanent magnets (Ostertag and Strnat, 1966; Ray and Strnat, 1972; Ojima et al., 1977). Following the development

DOI: 10.1201/9781003216346-8

of SmCo magnets heavily dependent on the expensive Sm/Co, technological efforts focus on the Fe-based hard magnetic alternatives, which led to the discovery of the third-generation NdFeB magnets in the early 1980s (Sagawa et al., 1984; Croat et al., 1984; Hadjipanayis et al., 1984). Owing to the high thermal stability over a broad operating temperature range, $SmCo_5$ and Sm_2Co_{17} permanent magnets are reliable for high-temperature applications up to 550 °C where material cost is not a limiting factor, including the safety-critical sectors like aerospace, and high-value sectors such as electric motors, generators and transformers (Strnat, 1978; Gutfleisch et al., 2006).

6.2 1:5-TYPE PERMANENT MAGNETS

A wide variety of intermetallic SmCo compounds exist due to the large size discrepancy between the RE element Sm and the TM element Co. The SmCo equilibrium phase diagram in Figure 6.1 displays the entire range of SmCo metallic compounds, including Sm_3Co, Sm_9Co_4, $SmCo_2$, $SmCo_3$, Sm_2Co_7, $SmCo_5$ and Sm_2Co_{17}, with their crystal structures summarized in Table 6.1 (Bushow and Van Der Goot, 1968). Among these phases, the

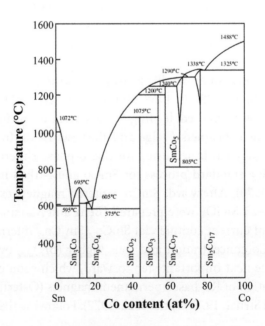

FIGURE 6.1 SmCo phase diagram. Reprinted with permission from (Bushow and Van Der Goot, 1968). Copyright (1968) Elsevier.

TABLE 6.1 Crystal Parameters of the SmCo Intermetallic Compounds (Bushow and Van Der Goot, 1968)

Compounds	Lattice Parameters	Crystal Symmetry	Structure Type
Sm_3Co	$a = 7.090$ Å $b = 9.625$ Å $c = 6.342$ Å	Orthorhombic	Fe_3C
Sm_9Co_4	$a = 11.15$ Å $b = 9.461$ Å $c = 9.173$ Å	Orthorhombic	Unknown
$SmCo_2$	$a = 7.260$ Å	Cubic	$MgCu_2$
$SmCo_3$	$a = 5.050$ Å $c = 24.590$ Å	Rhombohedral	$GdCo_3$
Sm_2Co_7	$a = 5.041$ Å $c = 24.327$ Å	Hexagonal	Ce_2Ni_7
$SmCo_5$	$a = 5.002$ Å $c = 3.964$ Å	Hexagonal	$CaZn_5$
Sm_2Co_{17} (R)	$a = 8.395$ Å $c = 12.216$ Å	Rhombohedral	Th_2Zn_{17}
Sm_2Co_{17} (H)	$a = 8.360$ Å $c = 8.515$ Å	Hexagonal	Th_2Ni_{17}

1:5-type and 2:17-type compounds with preferably high magnetization and Curie temperature are pursued as promising permanent magnets.

Figure 6.2 displays the $CaCu_5$-type crystal structure of $SmCo_5$ compound (*P6/mmm* space group, $a = 5.002$ Å, $c = 3.964$ Å). The unit cell contains one Sm site (1a) and two Co sites (2c, 3g). The base layer consists of both Sm (1a) and Co (2c) atoms, whereas the second layer consists exclusively of Co atoms at the 3g sites (Kumar, 1988). The Sm sublattice anisotropy is responsible for the large uniaxial anisotropy of $SmCo_5$ along the preferred crystallographic direction of hexagonal *c* axis, leading to a highest H_A of ~400 kOe that outperforming other $RECo_5$ systems (Croat, 2018). The M_s of the $SmCo_5$ is ~1.1 T, which is mainly attributed to the moment-bearing Co atoms (Harris and Jewell, 2012). The high T_C of ~1020 K is resulted from the strong Co–Co exchange interactions of the $SmCo_5$ lattice (Ucar et al., 2020).

Approaches to fabricate the $SmCo_5$ magnets include casting, melt-spinning, reduction diffusion, hot isostatic pressing and powder metallurgy (Szmaja, 2007; Li et al., 2012; An et al., 2019), with the main challenge of

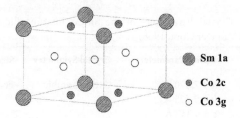

FIGURE 6.2 Hexagonal unit cell of the SmCo$_5$ intermetallic phase.

achieving high coercivity. In the ideal case, the coercivity reaches the upper limit of intrinsic H_A. However, for the nucleation-controlled SmCo$_5$ magnets, the practical H_{cj} measured at room temperature are less than 10–15% of the theoretical H_A. One common request for the high-coercivity SmCo$_5$ magnet is to minimize the number of nucleation sites of reversal domain by microstructure modification. For powder metallurgy as the most commonly used technique, green compacts with aligned magnetic powders are sintered at a high temperature (~1150 °C), subsequently slow cooling to ~900 °C for annealing, followed by quenching to room temperature to smooth the grain boundaries and hence increase the coercivity. As reported, the H_{cj} could be raised to ~3200 kA/m compared with ~80 kA/m without slow cooling (Harris and Jewell, 2012). Besides, excessive Sm has been introduced to restrain the formation of the unfavorable Sm$_2$Co$_{17}$ phase (Oesterreicherer, 1984). A certain amount of low-melting-point Sm-rich phase at the grain boundaries has also been introduced to promote the liquid phase sintering for full densification (Benz and Martin, 1972).

Investigations on the SmCo$_5$ magnets mainly focus on the following three aspects. Firstly, elemental substitution for either Sm or Co, and corresponding magnetic responses have been extensively investigated by experimental studies and density-functional-theory (DFT) calculations. Appropriate Tb/Dy/Er/Pr substitution for Sm in the SmCo$_5$ prototype has been reported to improve the M_s (Martis et al., 1978; Xu et al., 2016). For example, the Pr substitution for Sm in sintered magnet yields a high $(BH)_{max}$ of ~25 MGOe (Velu et al., 1989). Calculation studies have revealed that the common dopant Fe for Co also enhances the M_s of the SmCo$_{5-x}$Fe$_x$ on condition that the Fe content x ≤ 4 (Larson et al., 2004). The H_A can be simultaneously improved with the Fe content x ≤ 0.15, and then follows an unfortunately rapid decay with excessive Fe substitution (Larson et al., 2004). Introduction of Sn can form the nonmagnetic Sn-rich grain boundary (GB) phase, which enhances the coercivity of SmCo$_5$ melt-spun ribbons to 32 kOe (Kündig et al., 2006). However, the Si substitution for Co at both the 2c and 3g sites substantially deteriorates the uniaxial magnetic anisotropy and the magnetic moment (Chouhan et al., 2021). Other elements including Ni, Mn, Ti, Zr and Hf are thermodynamic stabilizers to form the 1:5-type structure whereas the Cr is detrimental to the phase stability (Söderlind et al., 2017; Landa et al., 2018; Mao et al., 2019).

Secondly, new routines have been applied to fabricate high-performance SmCo$_5$ magnets. For instance, low-temperature hot-pressing and hot-deformation have been explored to fabricate the nanostructured SmCo$_5$/Sm$_2$Co$_7$ composites with good texture along c axis, yielding high H_{cj} of 25.6 kOe and excellent temperature coefficient β of −0.23%/°C in the temperature range of 25–300 °C (Xu et al., 2020). Liquid-solution synthesis combined with reductive annealing has been used to synthesize well-oriented single-crystalline SmCo$_5$ particles for enhanced performance (Ma et al., 2016; Yue et al., 2018). In particular, Ma et al. have achieved the uniform single-crystalline SmCo$_5$ particles (~130 ± 10 nm) by designing a Co$_3$O$_4$@Sm$_2$O$_3$-CaO precursor, followed by dispersion in ethanol and mixing with epoxy to form the anisotropic SmCo$_5$ nanomagnets (Ma et al., 2019). This gives rise to large H_{cj} of 30.5 kOe, $(BH)_{max}$ of 18.1 MGOe and M_r/M_s of 0.95.

Thirdly, rational design of exchange-coupled SmCo$_5$/Co and SmCo$_5$/α-Fe nanocomposites has been proposed, combining the high coercivity from the nano-sized hard phase with high magnetization from the soft phase (Li et al., 2019; Kim et al., 2021; Xu et al., 2021). To obtain the homogeneous bi-anisotropic SmCo$_5$/Co nanocomposites, a simple chemical deposition approach has been taken, involving 60 wt% hard magnetic SmCo$_5$ nanoflakes and 40 wt% soft magnetic Co nanowires to generate a large $(BH)_{max}$ of 23.8 MGOe (Li et al., 2019). In the design of SmCo$_5$/α-Fe composite, through choosing the α-Fe nanoflakes to replace the traditional micron-scale α-Fe particles, the optimized microstructure and strong exchange coupling generates an increased $(BH)_{max}$ by 80% (Xu et al., 2021).

6.3 2:17-TYPE PERMANENT MAGNETS

Compared with the SmCo$_5$ that consumes a larger amount of scarce Sm, the Sm$_2$Co$_{17}$ compound possesses lower material cost, even higher M_s of ~1.2 T, theoretical $(BH)_{max}$ of ~39 MGOe, T_C of ~1200 K and sufficient H_A of ~60 kOe, triggering extensive research interest (Zhang et al., 2015;

Kronmüller and Goll, 2003; Gopalan et al., 2005). Although the Sm_2Co_{17} exhibits a larger M_s than that of the $SmCo_5$ due to the existence of increased Co proportion, it is rather difficult to achieve high H_{cj} in either the binary Sm_2Co_{17} or the isomorphic ternary $Sm_2(Co, Fe)_{17}$ alloys (Ostertag and Strnat, 1966; Ray and Strnat, 1972). Rational composition design via the addition of Fe, Cu and other refractory metallic elements such as Nb, V, Ta, Zr, etc. is necessary for successfully developing commercial 2:17-type magnets. Since the discovery of quinary $Sm_2(Co, Cu, Fe, Zr)_{17}$ with high $(BH)_{max}$ of 30 MGOe by Ojima et al., it has evolved into the strongest high-temperature hard magnetic materials over the past half-century (Ojima et al., 1977; Yu et al., 2017).

6.3.1 Phase Constituents

Current commercial $Sm_2(Co, Cu, Fe, Zr)_{17}$ mainly consists of three phases, including Sm_2Co_{17}, $SmCo_5$ and Zr-rich phase (Lucas et al., 2015; Gutfleisch et al., 2006). The Sm_2Co_{17} phase as the matrix phase determines the intrinsic magnetic properties. It possesses the rhombohedral Th_2Zn_{17} structure (space group $R\bar{3}m$, hereafter denoted as 2:17R) at room temperature, and transforms into the hexagonal Th_2Ni_{17} structure (space group $P6_3/mnc$, hereafter denoted as 2:17H) at elevated temperatures (Kumar, 1988; Harris and Jewell, 2012). Figure 6.3a shows the unit cell of Th_2Zn_{17}-type Sm_2Co_{17} phase, which consists of a six-layer repeating unit with one Sm site (6c) and four crystallographically non-equivalent Co sites (6c, 9d, 18j and 18h). Actually, the Sm_2Co_{17} structure is derived through ordered substitution of each third Sm atoms by a pair of Co atoms in the basal plane of the $SmCo_5$ structure. Figure 6.3b show the unit cell of Th_2Ni_{17}-type Sm_2Co_{17} phase,

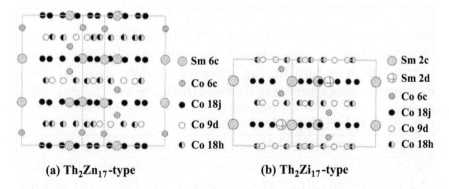

FIGURE 6.3 Unit cells of the (a) Th_2Zn_{17}-type and (b) Th_2Ni_{17}-type Sm_2Co_{17} phases.

including two Sm sites (2c, 2d) and four crystallographically non-equivalent Co sites (6c, 9d, 18j and 18h). The SmCo$_5$ with hexagonal CaCu$_5$ structure (hereafter abbreviated as 1:5H) is the cell boundary phase, which behaves as pinning sites during magnetization reversal, contributing to the coercivity enhancement (Romero et al., 2020). The Zr-rich lamellar phase can stabilize the cellular network, and simultaneously provide diffusion pathways for the Cu/Fe/Co atoms between cell interiors and boundaries, thereby modifying phase-ordering kinetics (Gutfleisch et al., 2006; Kronmüller and Goll, 2003). The understanding of such structure, however, has been controversial in the past few decades due to its limited thickness of 1–3 nm. Recently, through combination of advanced microstructural analysis, such as three-dimensional atom probe, nano-beam electron diffraction, and high-resolution transmission electron microscopy (HRTEM), it has been well identified that the Zr-rich phase is rhombohedral NbBe$_3$-type (Sm, Zr)(Co, Fe)$_3$ platelet and hereafter abbreviated as 1:3R (Xiong et al., 2004).

6.3.2 Fabrication Procedures

Multiple experimental studies on the Sm$_2$(Co, Cu, Fe, Zr)$_{17}$ have revealed that subtle microstructural changes may exert profound influences on the domain wall pinning and the final coercivity (Song et al., 2021). Consequently, the fabrication of 2:17-type magnets involves complicated and protracted heat treatment, with respective to that for the 1:5-type and 2:14:1-type magnets. As shown in Figure 6.4, the aligned and pressed 2:17-type compacts are sintered at ~1200 °C (1–3 h), and then subjected to solution treatment at 1100–1280 °C (4–10 h) before quickly quenching to room temperature to form the hexagonal TbCu$_7$-type solution precursor (1:7H phase) (Croat, 2018; Zhou et al., 2020). Next, the supersaturated 1:7H precursor is annealed at 750–890 °C (6–40 h), during which 1:7H phase decomposes to form the cellular nanostructure and Zr-rich lamella (Xiong et al., 2004). Finally, the magnets subjected to isothermal aging are slowly cooled at 0.4–1.0 °C/min to 400–500 °C, and annealed for 1–4 h. During this process, Cu and Fe redistribute between the SmCo$_5$ and Sm$_2$Co$_{17}$ phases (Gutfleisch et al., 2006). As illustrated in Figure 6.5a, the cellular structure consists of the 2:17R cell interiors indicated by A, and 1:5H cell boundary along pyramidal planes as strong pinning sites indicated by B, which is crucial to impede the domain wall motion for enhanced coercivity (Gutfleisch et al., 2006). As manifested in Figure 6.5b, the lamellar structure consists of elongated and thin 1:3R Zr-rich platelets extending over 2:17R cells. White arrow demonstrates that

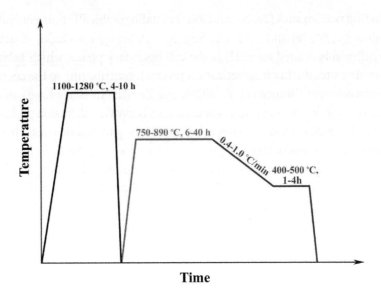

FIGURE 6.4 Schematic heat treatment regime for the Sm_2Co_{17}-type sintered magnets. Reprinted with permission from (Gutfleisch et al., 2006). Copyright (2006) Elsevier.

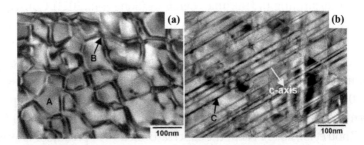

FIGURE 6.5 Bright-field TEM images of a $Sm_2(Co, Cu, Fe, Zr)_{17}$ sintered magnet with the nominal c axis of the 2:17R main phase (a) perpendicular and (b) parallel to the imaging plane (A: 2:17R cell interior, B:1:5H cell boundary phase, C: 1:3R Zr-rich phase). Reprinted with permission from (Gutfleisch et al., 2006). Copyright (2006) Elsevier.

the 1:3R Zr-rich phase (indicated by C) distributes parallel to the basal plane of 2:17R phase. Besides the commercial powder metallurgy approach with complicated post-sintering heat treatments, alternative melt-spinning route has also been established to develop the high-coercivity Sm_2Co_{17}-type magnets (Yan et al., 2005).

6.3.3 Latest Developments

Coercivity development of the 2:17-type permanent material follows two directions, including optimization of composition design and heat treatment, with the common goal of developing stronger pinning network. Influences of the Sm/Fe/Cu/Zr content on the magnetic properties and magnetization reversal process have been extensively investigated. In terms of the Sm(Co, Fe, Cu, Zr)$_z$ magnets, the increasing z value from 7.0 to 8.5 enlarges the average cell size from 88 to 108 nm, and enhances the H_{cj} from 15 to 35 kOe (Hadjipanayis et al., 2000). However, further increment of z to 9.1 results in the deficiency of 1:5 cell boundary phase and drastically reduces the H_{cj} to 8.0 kOe. The Fe-rich Sm_2Co_{17} magnets exhibit great potential toward low cost and high energy product (Song et al., 2020). Theoretically, higher Fe substitution content for Co permits larger M_s and resultantly enhanced $(BH)_{max}$. However, the appropriate Fe content to achieve the optimum coercivity is limited between 15–20 at%, as shown in Figure 6.6a (Tang et al., 2000). Raised Fe content above the critical value of 20 at% substantially lowers the density of 1:5H cell boundary phase, which is insufficient to form the complete cellular structure, as identified from the comparative Figures 6.6b and c. Increasing Fe content has been reported to promote the 1:7H → 2:17R transformation with Zr-rich 1:3R precipitates at the solution-treated sate, generating uneven cellular structures (Zhang et al., 2018a). It has also been reported that increasing Fe content yields larger 2:17R twin variants inside the cells and stacking faults at the cell edges, leading to less nucleus of the 1:5H cell boundaries. Meanwhile, the 2:7R, 1:3R, and 5:19R phases or their mixtures form at grain boundaries and generate the sparse 1:5H precipitates, which weaken the pinning strength for reduced coercivity (Song et al., 2020). Experimental results by Duerrschnabel et al. confirm the drastically deteriorated H_{cj} from 29.9 to 3.5 kOe with the Fe content increasing from 19 to 23 at%, accompanied with simultaneously reduced $(BH)_{max}$ from 32.9 to 12.6 MGOe (Duerrschnabel et al., 2017). To solve the above bottleneck concern, new approaches have been developed recently, including intermediate heat treatment (Horiuchi et al., 2015), pre-aging (Jia et al., 2020), multi-stage aging (Song et al., 2017b), and dual-alloying (Yan et al., 2019) to optimize the cellular nanostructure. Based on these advances, high $(BH)_{max}$ value above 32 MGOe and large H_{cj} over 30 kOe have been achieved. (Song et al., 2017a).

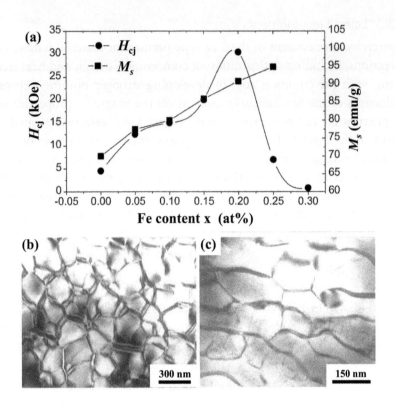

FIGURE 6.6 (a) Dependences of the H_{cj} and M_s on the Fe content of the Sm(Co$_{ba_1}$Fe$_x$Cu$_{0.128}$Zr$_{0.02}$)$_{7.0}$ sintered magnets. TEM images of the magnets with Fe contents of (b) $x = 0.20$ and (c) $x = 0.25$. Reprinted with permission from (Tang et al., 2000). Copyright (2000) Elsevier.

The Cu exhibits high solubility in the 1:5H phase compared to the 2:17R cell interiors. However, the concentration of Cu at GBs is lower than that inside grains, as typically shown in Figure 6.7 (Horiuchi et al., 2015). Such a Cu-lean phenomenon leads to incomplete cellular structure close to the grain boundary, unlike the uniform cellular structure inside the grains. Consequently, high coercivity value by increased Cu concentration in the Sm(Co, Fe, Cu, Zr)$_z$ is an important research direction. Most Cu-lean regions disappear, which generates better cellular structure and strengthens the pinning force for enhanced coercivity (Gopalan et al., 2009; Wang et al., 2018). Chen et al. have investigated the origin of the giant coercivity increment in 1 wt% Cu-particle-alloyed Sm(Co, Cu, Fe, Zr)$_7$, and found that Cu addition improves the continuity of 1:5H cell boundary with higher

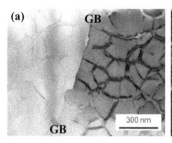 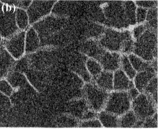

FIGURE 6.7 The Cu-lean phenomenon at the grain boundary of the aging-treated Sm(Co$_{bal.}$Fe$_{0.35}$Cu$_{0.06}$Zr$_{0.018}$)$_{7.8}$ magnet. (a) Bright-field TEM image, (b) energy-dispersive X-ray spectroscopy mapping of Cu. Reprinted with permission from (Horiuchi et al., 2015). Copyright (2015) AIP Publishing.

Cu content, facilitating the transformation of pinning mechanism from weaker repulsive pinning to stronger attractive pinning (Chen et al., 2019).

Trace Zr addition is a keystone for forming the cellular microstructure and improving the magnetic performance of 2:17-type magnets. For the Sm(Co$_{bal}$Cu$_{0.062}$Fe$_{0.285}$Zr$_x$)$_{7.6}$ magnets, increasing Zr content x from 0.016 to 0.023 promotes gradually increased density of the lamellar phase and increased Cu concentration in cell boundary phase for enhanced coercivity (Wang et al., 2020). Raised Zr content may also be beneficial to stabilize the 1:7H phase since ideal 1:7H single-phase precursor is the prerequisite for desired cellular microstructure after aging (Derkaoui et al., 1996; Li et al., 2018). Besides the 1:3R Zr-rich phase, excessive Zr addition also leads to the formation of soft magnetic 6:23 particles, as shown in Figure 6.8a–d (Yuan et al., 2020). From the high-magnification TEM image in Figure 6.8e, the 1:5H cell boundary phase is relatively rare around the 6:23 particles, which becomes local weak sites for pinning the domain wall motion and yields a poor squareness factor.

Great research interest also lies in the heat treatment to modify the cell structure toward high-performance Sm$_2$Co$_{17}$-type magnets, mainly involving the solution temperature, aging temperature and slow cooling process. As reported, the 1:7H single-phase can only be obtained at a proper solution temperature of ~1200 °C, while lower or higher solution temperatures deteriorate the magnetic properties due to the formation of FeCoZr and SmCoZr phases (Zhang et al., 2018b). Wang et al. have found that increasing aging temperature from 790 to 830 °C leads to the increment of cell size and the lamellar phase density, contributing to enhanced

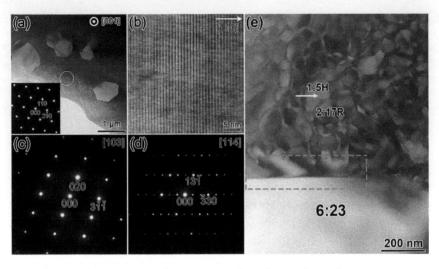

FIGURE 6.8 The formation of $Zr_6(Co, Fe)_{23}$ (6, 23) particles in the $Sm_{25}Co_{49.2}Fe_{16.2}Cu_{5.6}Zr_{4.0}$ sintered magnets. (a) Bright-field TEM image taken along $[001]_{2:17R}$ zone axis. White circle indicates where the inset diffraction pattern was taken. (b)–(d) HRTEM image and selected area electron diffraction patterns of the 6:23 Zr-rich particle from the red circle region in (a). (e) Enlarged view of the 1:5H precipitate free zone around the 6:23 Zr-rich particle. Reprinted with permission from (Yuan et al., 2020). Copyright (2020) Elsevier.

coercivity from 5.7 to 28.9 kOe (Wang et al., 2020). Further raising the aging temperature to 870 °C leads to uneven Cu concentration in cell boundaries and incomplete cell nanostructure. Sepehri-Amin et al. have compared the effect of different cooling processes on the magnetic performance of the sintered $Sm(Co, Fe, Cu, Zr)_{7.19}$, and reported that slow cooling yields a sharp and defect-free $SmCo_5/Sm_2Co_{17}$ interface (Sepehri-Amin et al., 2017). Large amounts of Cu enrichment in such sharp interface results in large H_A gradient from 2:17/1:5 interface to 1:5 cell boundary, which strengthens the pinning force for significant coercivity enhancement.

Another hot topic is the formation mechanism of the unique cell nanostructure and elemental segregation of 2:17-type sintered magnet. There has long been an open debate between two distinct models, i.e. pure atomic diffusion and displacive transformation. For the former model proposed by Livingston and Martin in 1977, the 2:17R phase precipitates and grows, while 1:5H phase remains from the 1:7H precursor (Livingston and Martin, 1977). The latter one involves the precipitation of 1:5H phase (Ray,

1990) and transformation from 2:17H to 2:17R via dislocation gliding (Rabenberg et al., 1982). A recent atomic-scale study on the microstructural evolution combined with microchemistry suggests that formed cellular nanostructure involves a diffusion-controlled displacive phase transformation (Song et al., 2020). During the aging of a supersaturated Sm(Co, Fe, Cu, Zr)$_{7.5}$ alloy, concurrent recrystallization and precipitation have been identified to originate from the progressive formation and dissociation of defects (stacking faults, vacancies and excessive interstitials). The diffusion-controlled dislocation glides transform the matrix from the 1:7H and 2:17H mixtures to Sm-depleted 2:17R cells, and offer continuous diffusion channels to form Sm-rich 1:5H cell boundary and Zr-rich 1:3R platelet. Another atomic-scale structural study also reveals the collective glide of three partial dislocations on successive basal planes to induce an ordered transformation from short-range 2:17R microtwins to long-range 2:17R phase during aging, together with the segregated Cu/Fe atoms in the 1:5/2:17 phases (Wu et al., 2020). These new findings by revisiting the evolution of cellular nanostructure are believed to promote further developments of 2:17-type hard magnetic materials.

REFERENCES

An, S. Z., W. H. Li, K. X. Song, T. L. Zhang, and C. B. Jiang. 2019. Phase transformation in anisotropic nanocrystalline SmCo$_5$ magnets. *Journal of Magnetism and Magnetic Materials* 469:113–8.

Benz, M. G., and D. L. Martin. 1970. Cobalt-samarium permanent magnets prepared by liquid phase sintering. *Applied Physics Letters* 17(4):176–9.

Benz, M. G., and D. L. Martin. 1972. Mechanism of sintering in cobalt rare-earth permanent-magnet alloys. *Journal of Applied Physics* 43(7):3165–70.

Buschow, K. H. J. 1988. Permanent magnet materials based on 3d-rich ternary compounds. In *Handbook of ferromagnetic materials*, ed. K. H. J. Buschow, 4:1–129. Amsterdam: North-Holland Publishing.

Bushow, K. H. J., and A. S. Van Der Goot. 1968. Intermetallic compounds in the system samarium-cobalt. *Journal of the Less Common Metals* 14(3):323–8.

Chen, H. S., Y. Q. Wang, Y. Yao, J. T. Qu, F. Yun, Y. Q. Li, S. P. Ringer, M. Yue, and R. K. Zheng. 2019. Attractive-domain-wall-pinning controlled Sm-Co magnets overcome the coercivity-remanence trade-off. *Acta Materialia* 164:196–206.

Chouhan, R. K., A. K. Pathak, and D. Paudyal. 2021. Understanding the origin of magneto-crystalline anisotropy in pure and Fe/Si substituted SmCo$_5$. *Journal of Magnetism and Magnetic Materials* 522:167549.

Croat, J. J. 2018. The development of rare earth permanent magnets. In *Rapidly solidified neodymium-iron-boron permanent magnets*, ed. J. J. Croat, 1–33. Cambridge: Woodhead Publishing.

Croat, J. J., J. F. Herbst, R. W. Lee, and F. E. Pinkerton. 1984. Pr-Fe and Nd-Fe-based materials: a new class of high-performance permanent magnets. *Journal of Applied Physics* 55(6):2078–82.

Derkaoui, S., N. Valignat, and C. H. Allibert. 1996. Co corner of the system Sm-Co-Zr: decomposition of the phase 1:7 and equilibria at 850°C. *Journal of Alloys and Compounds* 235:112–9.

Duerrschnabel, M., M. Yi, K. Uestuener, M. Liesegang, M. Katter, H. J. Kleebe, B. Xu, O. Gutfleisch, and L. Molina-Luna. 2017. Atomic structure and domain wall pinning in samarium-cobalt-based permanent magnets. *Nature Communications* 8:54.

Gopalan, R., D. H. Ping, K. Hono, M. Q. Huang, B. R. Smith, Z. M. Chen, and B. M. Ma. 2005. Investigation on structure-magnetic property correlation in melt-spun Sm(Co$_{0.56}$Fe$_{0.31}$Cu$_{0.04}$Zr$_{0.05}$B$_{0.04}$)$_z$ ribbons. *Journal of Magnetism and Magnetic Materials* 292:150–8.

Gopalan, R., K. Hono, A. Yan, and O. Gutfleisch. 2009. Direct evidence for Cu concentration variation and its correlation to coercivity in Sm(Co$_{0.74}$Fe$_{0.1}$Cu$_{0.12}$Zr$_{0.04}$)$_{7.4}$ ribbons. *Scripta Materialia* 60(6):764–67.

Gutfleisch, O., K. H. Müller, K. Khlopkov, M. Wolf, A. Yan, R. Schäfer, T. Gemming, and L. Schultz. 2006. Evolution of magnetic domain structures and coercivity in high-performance SmCo 2:17-type permanent magnets. *Acta Materialia* 54(4):997–1008.

Hadjipanayis, G. C., R. C. Hazelton, and K. R. Lawless. 1984. Cobalt-free permanent magnet materials based on iron-rare-earth alloys. *Journal of Applied Physics* 55(6):2073–77.

Hadjipanayis, G. C., W. Tang, Y. Zhang, S. T. Chui, J. F. Liu, C. Chen, and H. Kronmüller. 2000. High temperature 2:17 magnets: relationship of magnetic properties to microstructure and processing. *IEEE Transactions on Magnetics* 36:3382–7.

Harris, I. R., and G. W. Jewell. 2012. Rare-earth magnets: properties, processing and applications. In *Functional materials for sustainable energy applications*, ed. J. A. Kilner, S. J. Skinner, S. J. C. Irvine, and P. P. Edwards, 600–39. Cambridge: Woodhead Publishing.

Hoffer, G., and K. J. Strnat. 1966. Magnetocrystalline anisotropy of YCo$_5$ and Y$_2$Co$_{17}$. *IEEE Transactions on Magnetics* 2(3):487–9.

Horiuchi, Y., M. Hagiwara, M. Endo, N. Sanada, and S. Sakurada. 2015. Influence of intermediate-heat treatment on the structure and magnetic properties of iron-rich Sm(CoFeCuZr)$_z$ sintered magnets. *Journal of Applied Physics* 117(17):17C704.

Jia W. T., X. L. Zhou, A. D. Xiao, X. Song, T. Yuan, and T. Y. Ma. 2020. Defects-aggregated cell boundaries induced domain wall curvature change in Fe-rich Sm–Co–Fe–Cu–Zr permanent magnets. *Journal of Materials Science* 55(27):13258–69.

Kim, C. W., I. H. Kim, and Y. S. Kang. 2021. Magnetic spin exchange interaction in SmCo$_5$/Co nanocomposite magnet for large energy product. *Journal of Colloid and Interface Science* 589:157–65.

Kronmüller, H., and D. Goll. 2003. Analysis of the temperature dependence of the coercive field of Sm_2Co_{17} based magnets. *Scripta Materialia* 48:833–8.

Kumar, K. 1988. $RETM_5$ and RE_2TM_{17} permanent magnets development. *Journal of Applied Physics* 63(6):R13–57.

Kündig, A. A., R. Gopalan, T. Ohkubo, and K. Hono. 2006. Coercivity enhancement in melt-spun $SmCo_5$ by Sn addition. *Scripta Materialia* 54:2047–51.

Landa, A., P. Soderlind, D. Parker, D. Åberg, V. Lordi, A. Perron, P. E. A. Turchi, R. K. Chouhan, D. Paudyal, and T. A. Lograsso. 2018. Thermodynamics of $SmCo_5$ compound doped with Fe and Ni: an ab initio study. *Journal of Alloys and Compounds* 765:659–63.

Larson, P., I. I. Mazin, and D. A. Papaconstantopoulos. 2004. Effects of doping on the magnetic anisotropy energy in $SmCo_{5-x}Fe_x$ and $YCo_{5-x}Fe_x$. *Physical Review B* 69(13):134408.

Li, H. J., Q. Wu, M. Yue, L. Y. Cong, Y. T. ZhuGe, D. J. Wang, Y. Q. Li, and J. X. Zhang. 2019. A novel strategy for approaching high performance $SmCo_5$/Co nanocomposites. *Journal of Alloys and Compounds* 810:151890.

Li, T. Y., Z. Liu, Y. P. Feng, L. Liu, C. Y. Zhang, G. H. Yan, Z. X. Feng, D. Lee, and A. R. Yan. 2018. Effect of Zr on magnetic properties and electrical resistivity of $Sm(Co_{bal}Fe_{0.09}Cu_{0.09}Zr_x)_{7.68}$ magnets. *Journal of Alloys and Compounds* 753:162–6.

Li, W. F., H. Sepehri-Amin, L. Y. Zheng, B. Z. Cui, A. M. Gabay, K. Hono, W. J. Huang, C. Ni, and G. C. Hadjipanayis. 2012. Effect of ball-milling surfactants on the interface chemistry in hot-compacted $SmCo_5$ magnets. *Acta Materialia* 60:6685–91.

Livingston, J. D., and D. L. Martin. 1977. Microstructure of aged $(Co,Cu,Fe)_7Sm$ magnets. *Journal of Applied Physics* 48(3):1350–4.

Lucas, J., P. Lucas, T. L. Mercier, A. Rollat, and W. Davenport. 2015. Rare earth-based permanent magnets preparation and uses. *Rare Earths* 14:231–49.

Ma, Z. H., J. M. Liang, W. Ma, L. Y. Cong, Q. Wu, and M. Yue. 2019. Chemically synthesized anisotropic $SmCo_5$ nanomagnets with a large energy product. *Nanoscale* 11:12484–8.

Ma, Z. H., S. X. Yang, T. L. Zhang, and C. B. Jiang. 2016. The chemical synthesis of $SmCo_5$ single-crystal particles with small size and high performance. *Chemical Engineering Journal* 304:993–9.

Mao, F., H. Lu, D. Liu, K. Guo, F. W. Tang, and X. Y. Song. 2019. Structural stability and magnetic properties of $SmCo_5$ compounds doped with transition metal elements. *Journal of Alloys and Compounds* 810:151888.

Martis, R. J. J., N. Gupta, S. G. Sankar, and V. U. S. Rao. 1978. Temperature compensated magnetic materials of the type $Sm_xR_{1-x}Co_5$ (R = Tb, Dy, Er). *Journal of Applied Physics* 49(3):2070–1.

Oesterreicherer, H. 1984. On the coercivity of cellular Sm_2Co_{17}-$SmCo_5$ permanent magnets. *Journal of the Less Common Metals* 99:L17–20.

Ojima, S., S. Tomizawa, T. Yoneyama, and T. Hori. 1977. New type rare earth cobalt magnets with an energy product of 30 MG·Oe. *The Japan Society of Applied Physics* 16(4):671–2.

Ostertag, W., and K. J. Strnat. 1966. Rare earth cobalt compounds with the A_2B_{17} Structure. *Acta Crystallographica* 21(4):560–5.

Rabenberg, L., R. Mishra, and G. Thomas. 1982. Development of the cellular microstructure in the SmCo7.4-type magnets, in *Proc. 6th Inter. Workshop on REPM*, 599–608.

Ray, A. E. 1990. A revised model for the metallurgical behavior of 2:17 type permanent magnet alloys. *Journal of Applied Physics* 67(9):4972–4.

Ray, A. E., and K. J. Strnat. 1972. Easy directions of magnetization in Ternary $R_2(Co, Fe)_{17}$ Phases. *IEEE Transactions on Magnetics* 8(3):516–8.

Romero, S. A., A. J. Moreira, F. F. G. Landgraf, and M. F. Campos. 2020. Abnormal coercivity behavior and magnetostatic coupling in SmCoCuFeZr magnets. *Journal of Magnetism and Magnetic Materials* 514:167147.

Sagawa, M., S. Fujimori, M. Togawa, and Y. Matsuura. 1984. New material for permanent magnets on a base of Nd and Fe. *Journal of Applied Physics* 55(6):2083–7.

Sepehri-Amin, H., J. Thielsch, J. Fischbacher, T. Ohkubo, T. Schrefl, O. Gutfleisch, and K. Hono. 2017. Correlation of microchemistry of cell boundary phase and interface structure to the coercivity of $Sm(Co_{0.784}Fe_{0.100}Cu_{0.088}Zr_{0.028})_{7.19}$ sintered magnets. *Acta Materialia* 126:1–10.

Söderlind, P., A. Landa, I. L. M. Locht, D. Åberg, Y. Kvashnin, M. Pereiro, M. Däne, P. E. A. Turchi, V. P. Antropov, and O. Eriksson. 2017. Prediction of the new efficient permanent magnet $SmCoNiFe_3$. *Physical Review B* 96(10):100404.

Song, K. K., W. Sun, H. S. Chen, N. J. Yu, Y. K. Fang, M. G. Zhu, and W. Li. 2017a. Revealing on metallurgical behavior of iron-rich $Sm(Co_{0.65}Fe_{0.26}Cu_{0.07}Zr_{0.02})_{7.8}$ sintered magnets. *AIP Advances* 7(5):056238.

Song, K. K., Y. K. Fang, W. Sun, H. S. Chen, N. J. Yu, M. G. Zhu, and W. Li. 2017b. Microstructural analysis during the step-cooling annealing of iron-rich $Sm(Co_{0.65}Fe_{0.26}Cu_{0.07}Zr_{0.02})_{7.8}$ anisotropic sintered magnets. *IEEE Transactions on Magnetics* 53(11):2100804.

Song, X., T. Y. Ma, X. L. Zhou, F. Ye, T. Yuan, J. D. Wang, M. Yue, F. Liu, and X. B. Ren. 2021. Atomic scale understanding of the defects process in concurrent recrystallization and precipitation of Sm-Co-Fe-Cu-Zr alloys. *Acta Materialia* 202:290–301.

Song, X., X. L. Zhou, T. Yuan, J. D. Wang, M. Yue, and T. Y. Ma. 2020. Role of nanoscale interfacial defects on magnetic properties of the 2:17-type SmCo permanent magnets. *Journal of Alloys and Compounds* 816:152620.

Strnat, K. J. 1978. Rare-earth magnets in present production and development. *Journal of Magnetism and Magnetic Materials* 7(1–4):351–60.

Szmaja, W. 2007. Studies of the domain structure of anisotropic sintered $SmCo_5$ permanent magnets. *Journal of Magnetism and Magnetic Materials* 311(2):469–80.

Tang, W., Y. Zhang, and G. C. Hadjipanayis. 2000. Microstructure and magnetic properties of $Sm(Co_{bal}Fe_xCu_{0.128}Zr_{0.02})_{7.0}$ magnets with Fe substitution. *Journal of Magnetism and Magnetic Materials* 221(3):268–72.

Ucar, H., R. Choudhary, and D. Paudyal. 2020. An overview of the first principles studies of doped RE-TM$_5$ systems for the development of hard magnetic properties. *Journal of Magnetism and Magnetic Materials* 496:165902.

Velge, W. A. J. J., and K. H. J. Bushow. 1968. Magnetic and crystallographic properties of some rare earth cobalt compounds with CaZn$_5$ Structure. *Journal of Applied Physics* 39(3):1717–20.

Velu, E. M. T., R. T. Obermyer, S. G. Sankar, and W. E. Wallace. 1989. PrCo$_5$-based high-energy-density permanent magnets. *Journal of the Less Common Metals* 148:67–71.

Wang, S., Y. K. Fang, C. Wang, L. Wang, M. G. Zhu, W. Li, and G. C. Hadjipanayis. 2020. Dependence of macromagnetic properties on the microstructure in high-performance Sm$_2$Co$_{17}$-type permanent magnets. *Journal of Magnetism and Magnetic Materials* 510:166942.

Wang, Y. Q., M. Yue, D. Wu, D. T. Zhang, W. Q. Liu, and H. G. Zhang. 2018. Microstructure modification induced giant coercivity enhancement in Sm(CoFeCuZr)$_z$ permanent magnets. *Scripta Materialia* 146:231–5.

Wu, H. C., C. Y. Zhang, Z. Liu, G. Q. Wang, H. M. Lu, G. X. Chen, Y. Li, R. J. Chen, and A. R. Yan. 2020. Nanoscale short-range ordering induced cellular structure and microchemistry evolution in Sm$_2$Co$_{17}$-type magnets. *Acta Materialia* 200:883–92.

Xiong, X. Y., T. Ohkubo, T. Koyama, K. Ohashi, Y. Tawara, and K. Hono. 2004. The microstructure of sintered Sm(Co$_{0.72}$Fe$_{0.20}$Cu$_{0.055}$Zr$_{0.025}$)$_{7.5}$ permanent magnet studied by atom probe. *Acta Materialia* 52(3):737–48.

Xu, C., H. Wang, B. J. Liu, H. Xu, T. L. Zhang, J. H. Liu, and C. B. Jiang. 2021a. Correlation between ordered solid solution and cellular structure of Sm$_2$Co$_{17}$ type magnets with high iron content. *Journal of Magnetism and Magnetic Materials* 519:167477.

Xu, X. C., Y. Q. Li, Z. H. Ma, M. Yue, and D. T. Zhang. 2020. Sm$_2$Co$_7$ nanophase inducing low-temperature hot deformation to fabricate high performance SmCo$_5$ magnet. *Scripta Materialia* 178:34–8.

Xu, X. C., Y. Q. Li, Z. H. Ma, Y. T. ZhuGe, Y. Teng, Y. Zhang, H. G. Zhang, D. T. Zhang, and M. Yue. 2021. Tuning the morphology of soft magnetic phase to optimize the microstructure of SmCo$_5$/α-Fe nanocomposites. *Materials Characterization* 172:110838.

Yan, A., A. Handstein, T. Gemming, K. H. Müller, and O. Gutfleisch. 2005. Coercivity mechanism of Sm$_2$(Co, Cu, Fe, Zr)$_{17}$-based magnets prepared by melt-spinning. *Journal of Magnetism and Magnetic Materials* 290–291:1206–9.

Yan, G. H., Z. Liu, W. X. Xia, C. Y. Zhang, G. Q. Wang, R. J. Chen, D. Lee, J. P. Liu, and A. R. Yan. 2019. Grain boundary modification induced magnetization reversal process and giant coercivity enhancement in 2:17 type SmCo magnets. *Journal of Alloys and Compounds* 785:429–35.

Yu, N. J., M. G. Zhu, Y. K. Fang, L. W. Song, W. Sun, K. K. Song, Q. Wang, and W. Li. 2017. The microstructure and magnetic characteristics of Sm(Co$_{bal}$Fe$_{0.1}$Cu$_{0.09}$Zr$_{0.03}$)$_{7.24}$ high temperature permanent magnets. *Scripta Materialia* 132:44–8.

Yuan, T., X. Song, X. L. Zhou, W. T. Jia, M. Musa, J. D. Wang, and T. Y. Ma. 2020. Role of primary Zr-rich particles on microstructure and magnetic properties of 2:17-type Sm-Co-Fe-Cu-Zr permanent magnets. *Journal of Materials Science & Technology* 53:73–81.

Yue, M., C. L. Li, Q. Wu, Z. H. Ma, H. H. Xu, and S. Palaka. 2018. A facile synthesis of anisotropic $SmCo_5$ nanochips with high magnetic performance. *Chemical Engineering Journal* 343:1–7.

Zhang, C. Y., Z. Liu, M. Li, L. Liu, T. Y. Li, R. J. Chen, D. Lee, and A. R. Yan. 2018a. The evolution of phase constitution and microstructure in iron-rich 2:17-type Sm-Co magnets with high magnetic performance. *Scientific Reports* 8:9103.

Zhang, T. L., H. Y. Liu, J. H. Liu, and C. B. Jiang. 2015. 2:17-type SmCo quasi-single-crystal high temperature magnets. *Applied Physics Letters* 106(16):162403.

Zhang, T. L., Q. Song, H. Wang, J. M. Wang, J. H. Liu, and C. B. Jiang. 2018b. Effects of solution temperature and Cu content on the properties and microstructure of 2:17-type SmCo magnets. *Journal of Alloys and Compounds* 735:1971–6.

Zhou, X. L., X. Song, W. T. Jia, A. D. Xiao, T. Yuan, and T. Y. Ma. 2020. Identifications of $SmCo_5$ and $Sm_{n+1}Co_{5n-1}$-type phases in 2:17-type Sm-Co-Fe-Cu-Zr permanent magnets. *Scripta Materialia* 182:1–5.

FURTHER READING

Duerrschnabel, M., M. Yi, K. Uestuener, M. Liesegang, M. Katter, H. J. Kleebe, B. Xu, O. Gutfleisch, and L. Molina-Luna. 2017. Atomic structure and domain wall pinning in samarium-cobalt-based permanent magnets. *Nature Communications* 8:54.

Song, X., T. Y. Ma, X. L. Zhou, F. Ye, T. Yuan, J. D. Wang, M. Yue, F. Liu, and X. B. Ren. 2021. Atomic scale understanding of the defects process in concurrent recrystallization and precipitation of Sm-Co-Fe-Cu-Zr alloys. *Acta Materialia* 202:290–301.

CHAPTER 7

Rare-Earth-Based Hard Magnetic Materials: NdFeB

7.1 INTRODUCTION TO NdFeB

The discovery of $Nd_2Fe_{14}B$ ternary phase has long been desired by the hard magnetic material community, which is largely composed of the inexpensive Fe providing high magnetization, small amounts of rare earths (REs) affording the majority of anisotropy, and minor B stabilizing the tetragonal phase. Illustrated by the unit cell in Figure 7.1 (Harris and Jewell, 2012), it is constructed by four $Nd_2Fe_{14}B$ units with space group $P4_2/mnm$. 56 Fe atoms share six sites (4e, 4c, $8j_1$, $8j_2$, $16k_1$, $16k_2$), 8 Nd atoms occupy crystallographically non-equivalent 4f and 4g sites, while 4 B atoms are located at 4g site. The $Nd_2Fe_{14}B$ tetragonal phase determines the intrinsic magnetic parameters, including saturation magnetic polarization J_s, magnetocrystalline anisotropy field H_A and Curie temperature T_C. The practical magnetic properties, however, depend crucially on the multi-phase microstructure, i.e. $Nd_2Fe_{14}B$ matrix phase (dark contrast) separated by the continuous Nd-rich intergranular phase (bright contrast), as shown in the back-scattered scanning electron microscopic (SEM) image in Figure 7.2. The Nd-rich phase plays two predominant roles, i.e. to assist the liquid phase sintering due to its low melting point, and to isolate the short-range exchange-coupling between the neighboring ferromagnetic $Nd_2Fe_{14}B$ grains due to its non-ferromagnetism and uniform distribution.

DOI: 10.1201/9781003216346-9

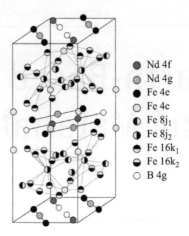

FIGURE 7.1 Unit cell of $Nd_2Fe_{14}B$ tetragonal compound. Reprinted with permission from (Harris and Jewell, 2012). Copyright (2012) Elsevier.

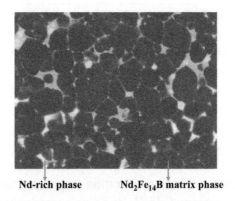

FIGURE 7.2 Typical back-scattered SEM image of the NdFeB sintered magnet.

7.2 RESEARCH FOCUSES OF NdFeB

Given the pivotal role of NdFeB magnet in a multitude of applications, its production has grown at an average annual rate of 10–20% for the past two decades, and is anticipated to experience steady prosperity in the future. This along comes a raising demand for indispensable REs. Table 7.1 summarizes the intrinsic magnetic properties, density and lattice parameters of $RE_2Fe_{14}B$ compound, including the most commonly used REs Nd/Pr/Dy/Tb, and the abundant REs La/Ce/Y (Herbst, 1991). Evidently, the $(Dy/Tb)_2Fe_{14}B$ compound possesses higher H_A than that of $(Nd/Pr)_2Fe_{14}B$,

TABLE 7.1 Intrinsic Magnetic Properties, Density and Lattice Parameters of Tetragonal $RE_2Fe_{14}B$ Compounds at Room Temperature (Herbst, 1991)

Compound	J_s (T)	H_A (kOe)	T_C (K)	ρ (g/cm³)	a (Å)	c (Å)
$La_2Fe_{14}B$	1.38	20	530	7.40	8.82	12.34
$Ce_2Fe_{14}B$	1.17	26	424	7.67	8.76	12.11
$Pr_2Fe_{14}B$	1.56	75	565	7.54	8.80	12.23
$Nd_2Fe_{14}B$	1.60	73	585	7.60	8.80	12.20
$Tb_2Fe_{14}B$	0.70	~220	620	7.96	8.77	12.05
$Dy_2Fe_{14}B$	0.71	~150	598	8.05	8.76	12.01
$Y_2Fe_{14}B$	1.41	26	565	7.00	8.76	12.00

being essential to prepare high-coercivity REFeB magnets. However, both Dy and Tb have been categorized as critical metals because of their geographic scarcity and extraction difficulty (Jin et al., 2016a). Meanwhile, the abundant and inexpensive REs including La, Ce and Y as by-products of the Nd/Pr/Dy/Tb during the mineral extraction, are massively backlogged due to the inferior magnetism of $(La/Ce/Y)_2Fe_{14}B$ compound to $Nd_2Fe_{14}B$ (Pathak et al., 2015). Taking the worldwide largest Bayan Obo RE ore as a typical example, abundant La/Ce account for ~73% of the total REs, while no more than 10% La/Ce have been utilized in REFeB. To tackle the global RE criticality and unbalanced RE utilization, tremendous efforts have been devoted to developing high-coercivity REFeB with less Dy/Tb, and low-cost REFeB with more La/Ce/Y.

7.2.1 High-Coercivity REFeB with Less Dy/Tb

For conventional NdFeB, the coercivity of approximately 1.2 T at room temperature is insufficient for applications in wind turbine generators and electric vehicle motors, due to the thermal demagnetization at high service temperatures above 200 °C. Theoretically, the coercivity H_{cj} can be enhanced intrinsically via partial replacement of Nd by the heavy REs Dy/Tb with large H_A, but at the cost of decreased remanence B_r and maximum energy product $(BH)_{max}$ since Dy/Tb and Fe atoms are antiferromagnetically coupled. Figure 7.3 shows a map of $(BH)_{max}$ vs. H_{cj} of commercial NdDyFeB sintered magnets with different Dy contents for specific applications (Oono et al., 2011). For hybrid electronic vehicle (HEV) motors with improved coercivity above 30 kOe, 10 wt% Dy substitution for Nd is necessary.

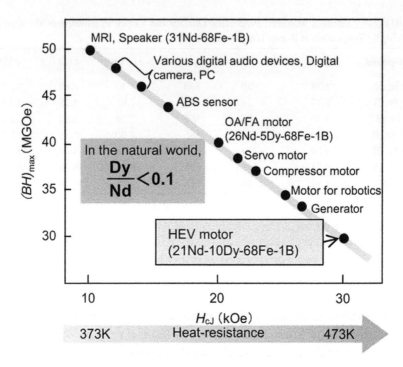

FIGURE 7.3 A map of the magnetic properties of commercially available NdDyFeB sintered magnets for specific applications. Reprinted with permission from (Oono et al., 2011). Copyright (2011) Elsevier.

To mitigate the excessive Dy/Tb consumption without sacrificing coercivity has been the pursuit of academia and industry in the past decades. The coercivity of sintered NdFeB is mainly determined by the nucleation of reversal magnetic domains at the epitaxial defect layer of 2:14:1 grains with locally lower H_A (Ramesh and Srikrishna, 1988; Fukuno et al., 1990). Accordingly, an approach named as grain boundary diffusion process (GBDP) has been proposed to enable more efficient Dy/Tb utilization (Hirota et al., 2006). During the GBDP, Dy/Tb-containing sources are coated on the magnet surface in the form of oxide, fluoride, pure metal or eutectic alloys, followed by heat treatment (usually below 1000 °C for 3–10 h) to allow Dy/Tb diffusion from the surface capping layer toward the inner regions. Figure 7.4a demonstrates typical Dy diffusion along the GBs, where the majority of Dy forms Dy-rich magnetically hardening shells surrounding the $Nd_2Fe_{14}B$ grains (Sugimoto, 2011), which is different from direct alloying with homogeneous Dy/Tb distribution inside the

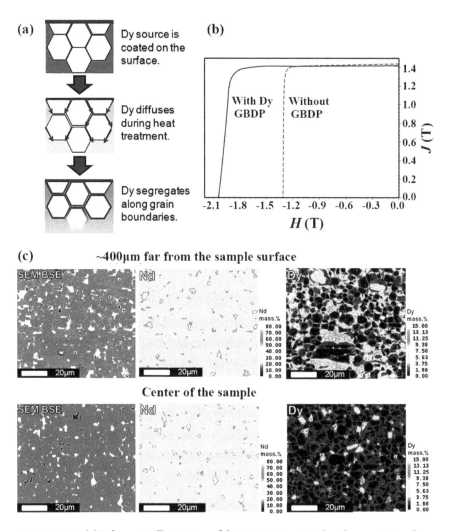

FIGURE 7.4 (a) Schematic illustration of the GBDP. Reprinted with permission from (Sugimoto, 2011). Copyright (2011) IOP Publishing. (b) Demagnetization curves of the NdFeB sintered magnets with or without Dy GBDP, (c) back-scattered SEM image and elemental mapping from surface and center of the GBDP sample. Reprinted with permission from (Sepehri-Amin et al., 2013). Copyright (2013) Elsevier.

grains (Li et al., 2011). Figure 7.4b displays the effects of GBDP using Dy vapor deposition, which increases the coercivity from 1.31 to 2.04 T, accompanied with slight remanence reduction from 1.44 to 1.42 T (Sepehri-Amin et al., 2013). For the sake of more efficient coercivity increment at minimized Dy/Tb utilization, different diffusion sources through sputtering, vapor deposition, coating and soaking have been designed, including DyNiAl, DyCu, PrTbCuAl, TbH$_x$ and TbCu, etc. (Oono et al., 2011; Chen et al., 2018; Kim et al., 2019; Liu et al., 2020a; Zhao et al., 2020; Sodernik et al., 2021; Lu et al., 2021). Influences of diffusion directions either perpendicular or parallel to the magnetically easy c-axis have also been compared (Ma et al., 2015). Despite such efforts, since the diffusion distance of Dy/Tb is limited under certain diffusion time and temperature, both the thickness and Dy/Tb concentration of the (Nd, Dy/Tb)$_2$Fe$_{14}$B shell decrease from the surface to the center region (Figure 7.4c), restraining the thickness of magnet subjected to the GBDP treatment (usually below 10 mm).

Another novel approach to develop high-coercivity REFeB with less Dy/Tb is grain boundary restructuring (GBR), which involves designing new Dy/Tb-rich GB alloy to replace the Nd-rich phase, and tuning the processing parameters to achieve required microstructures (Ni et al., 2012). The Dy/Tb-rich GB alloy and the near-stoichiometric 2:14:1 matrix alloy are separately smelted, pulverized and mixed homogeneously at certain proportion, followed by magnetic alignment and compressing, sintering and annealing. Figure 7.5a–f show the microstructure and elemental distribution of the Dy$_{71.5}$Fe$_{28.5}$ GBR magnet as an example, where the 2:14:1 matrix grains with Dy-rich shells covering the Nd-rich cores have been achieved (Zhang et al., 2020). Being different from the discontinuous Nd-rich GBs, as well as rough Nd$_2$Fe$_{14}$B/Nd-rich interface with non-stoichiometric composition and locally low anisotropy acting as primary nucleation of reversal domain (Figures 7.5g and h), the restructured continuous GB phase and the magnetic hardening shell with higher H_A can impede the nucleation of reversal domain for enhanced coercivity at low Dy consumption (Figure 7.5i).

Rational design of new GB alloy is critical to the final performance of GBR magnet, for which well-defined guidelines have been established (Yan et al., 2019), including: (i) thermodynamically, the elements of GB alloy should hardly diffuse into the Nd$_2$Fe$_{14}$B matrix phase to guarantee stable 2:14:1 tetragonal structure and maintained intrinsic magnetic properties; (ii) no occurrence of soft magnetic phase based on the phase diagram of the GB alloy is necessary to suppress the magnetic dilution; and (iii) with the aim of high coercivity, heavy-REs-containing new GB phase with low melting point and satisfactory wettability are essential for achieving the

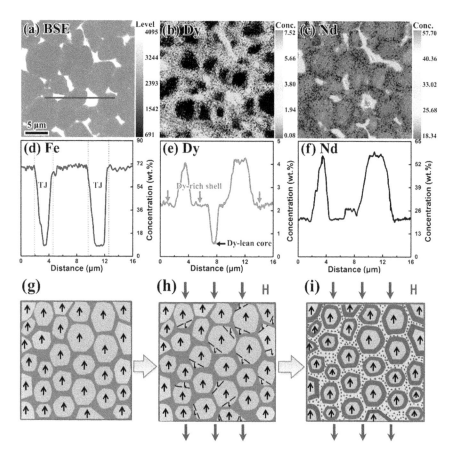

FIGURE 7.5 (a–f) Morphology and elemental distribution of 2 wt% $Dy_{71.5}Fe_{28.5}$ GBR magnet as a typical example. Reprinted with permission from (Zhang et al., 2020). Copyright (2020) Elsevier. (g–i) Schematic illustration of GBR approach, pristine magnet (g) before and (h) after applying a reversal magnetic field, (i) GBR magnet.

Dy/Tb-rich magnetic hardening shell, full densification and homogeneous microstructure during sintering.

Based on the above guidelines, a series of GB alloys have been employed to enhance the overall magnetic properties of REFeB magnets. Examples include $Dy_{69}Ni_{31}$ (Liu et al., 2014), $Dy_{32.5}Fe_{62}Cu_{5.5}$ (Liang et al., 2014), DyH_x (Liu et al., 2015), and $Dy_{71.5}Fe_{28.5}$ (Liang et al., 2015). Further modification of new intergranular phases via nanotechnology has also developed (Zhang et al., 2019). As depicted in Figure 7.6i, the nano powders in the GB regions can inhibit the growth of $Nd_2Fe_{14}B$ grains during sintering, and

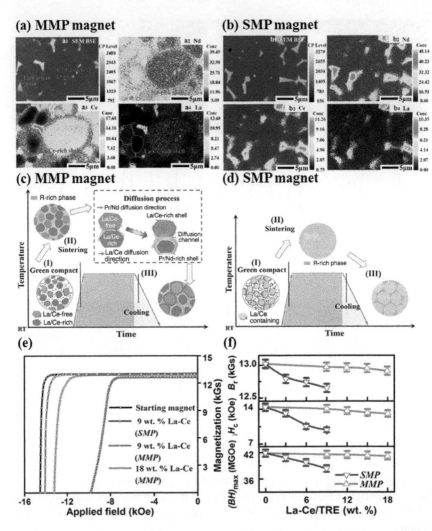

FIGURE 7.6 Back-scattered SEM images and corresponding elemental mappings for the (a) MMP and (b) SMP magnets. Schematics in (c) and (d) show the different microstructural evolution of MMP and SMP magnets. (e) Demagnetization curves and (f) dependence of magnetic properties on the La-Ce concentration for SMP and MMP magnets (Jin et al., 2016a). (Open access article.)

modify the distribution of the restructured GB phase toward thermally stable REFeB magnets for high-temperature applications up to 250 °C. The GBR route provides a rich spectrum of possibilities to fabricate high-coercivity 2:14:1-type hard magnetic materials at low Dy/Tb consumption.

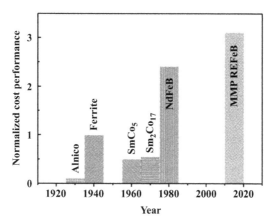

FIGURE 7.7 Comparison of the cost performance for the major technologically important permanent magnetic materials, including Alnico, ferrite, SmCo$_5$, Sm$_2$Co$_{17}$, NdFeB, and newly developed MMP REFeB magnets (Jin et al., 2016a). (Open access article.)

Since the GBR approach does not complicate the production process, it is highly potential for massive production as well.

Other ways toward Dy/Tb-free high-coercivity REFeB have also been explored, involving grain size reduction (Hu et al., 1999; Kim et al., 2011; Sepehri-Amin et al., 2011; Kobayashi et al., 2013) and non-ferromagnetic 6:13:1-type GB phase by Ga or Cu doping (Sasaki et al., 2016; Xu et al., 2018; Li et al., 2020; Fu et al., 2021), which are based on the manipulation of microchemistry and microstructure of GB phases. Detailed information can be referred to the book by Sepehri-Amin et al. (2018).

7.2.2 Low-Cost REFeB with More La/Ce/Y

Earlier investigations on the low-cost REFeB magnets have revealed that incorporating La/Ce/Y to substitute Nd/Pr significantly deteriorates the magnetic performance (Okada et al., 1985; Tang et al., 1989), stemming from the much lower J_s, H_A and T_C of (La/Ce/Y)$_2$Fe$_{14}$B compound than Nd$_2$Fe$_{14}$B (Table 7.1). Therefore, it is rather challenging to fabricate the high-performance La/Ce/Y-rich REFeB magnets. In the recent decade, a novel prototype of multi-main-phase (MMP) REFeB has been proposed with chemical heterogeneity (Jin et al., 2016a; Zhu et al., 2014), i.e. RE concentration varies significantly for the 2:14:1 grains. As shown in Figure 7.6a, the La/Ce-lean grain is covered by a thick La/Ce-rich shell, and the

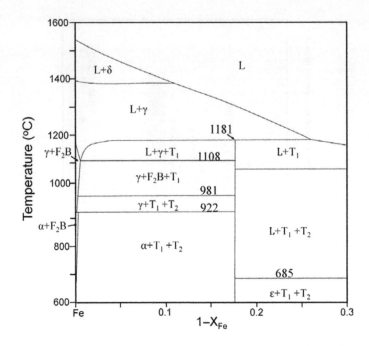

FIGURE 7.8 The Fe part of the NdFeB phase diagram with Nd:B = 2:1. T_1 and T_2 represent the $Nd_2Fe_{14}B$ phase and the $Nd_{1.1}Fe_4B_4$ phase, respectively. Reprinted with permission from (Hallemans et al., 1995). Copyright (1995) Springer Nature.

La/Ce-rich grain is covered by a Nd-rich shell. Differently, for the single-main-phase (SMP) magnet prepared via La/Ce directly alloying, La/Ce/Nd are homogeneously distributed within the matrix phases (Figures 7.6b and d). The distinct core-shell structure of MMP magnet stems from the giant composition difference between the initial La/Ce-rich and La/Ce-free components in the green compact, which is followed by the elemental inter-diffusion process during sintering (Figure 7.6c). As a result, the strong short-range exchange-coupling between core-shells of the 2:14:1 grain, and the long-range magnetostatic interaction across grains contribute to larger B_r, H_{cj} and $(BH)_{max}$ of the MMP magnets than those of the SMP magnets at an identical La/Ce substitution level (Figures 7.6e and f). The MMP strategy provides a promising solution not only to fabricate REFeB magnets with high cost performance, but also to the global REs criticality.

Since single La-substituted magnet faces the challenges of unstable 2:14:1 tetragonal phase and severely deteriorated microstructure, and

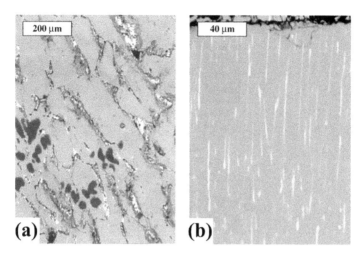

FIGURE 7.9 Back-scattered SEM images of the as-cast alloy taken from the (a) conventional ingot and (b) cast strip. Reprinted with permission from (Pei et al., 2002). Copyright (2002) Elsevier.

single Ce-substituted magnet exhibits poor thermal stability, a possible route lies in the introduction of La-Ce and Y-Ce RE mixtures to take full advantage of the synergetic effects between different REs. $La_{35}Ce_{65}$ co-substitution for $Nd_{80}Pr_{20}$ has been proved to stabilize the 2:14:1 tetragonal phase over a wide composition range, and to tailor the Ce valence toward the preferable trivalent state, which offers an effective approach to maintain the intrinsic magnetism (Jin et al., 2016b; Liu et al., 2021). In addition, the direct utilization of La-Ce raw materials with crustal ratio of 35:65 rather than pure La or Ce metal is environmental-friendly and cost-effective by skipping the multi-step extraction. The RE Y has also been found to enhance the thermal stability of Ce-rich REFeB magnets (Peng et al., 2017, 2020; Fan et al., 2018; Wu et al., 2021). This can result from the weakened temperature dependences of H_A and J_s as the Y preferably enters the 2:14:1 phase (Liu et al., 2022).

La-Ce and Y-Ce co-substitutions for Nd-Pr via the MMP approach allows the development of high-performance and low-cost La/Ce/Y-rich REFeB, acting as a part of endeavor for balanced utilization of strategic RE sources. From the viewpoint of fundamental research, it also opens up a new horizon on the synergistic effects of multi-RE substitution. In the context of high La/Ce/Y substitution above 30 wt%, we are only halfway to the high-performance REFeB commercial magnets. Based on the Object

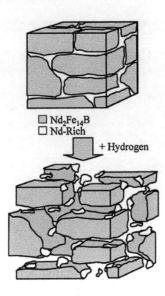

FIGURE 7.10 Schematic illustration of the HD process during the manufacturing of sintered NdFeB. Reprinted with permission from (Zakotnik et al., 2009). Copyright (2009) Elsevier.

Oriented Micro Magnetic Framework (OOMMF) micromagnetic simulation and in - situ Lorentz transmission electron microscopic characterization, the scenario to enhance the magnetic performance of MMP REFeB magnet is unveiled, i.e. forming nonmagnetic thick GB layer and retaining the chemical heterogeneity among main phases (Zhang et al., 2018; Jin et al., 2019a). Key technologies including low-temperature sintering (Jin et al., 2020), skipping the post-sinter annealing (Jin et al., 2019a), grain boundary restructuring and diffusion (Jin et al., 2021; Liu et al., 2020b) are proposed accordingly. The MMP REFeB are developing promisingly compared with other permanent magnets in terms of high-cost performance (Figure 7.7). Till now, the MMP REFeB have been successfully fabricated under industrial conditions (≥100 kg), with the La/Ce/Y substitution level as high as 50 wt% of the total REs and with comparable magnetic properties to commercial sintered NdFeB (Jin et al., 2022).

It is also worth noting that the cost-effective recycling of NdFeB is an important direction in the transition to the green economy (Yin et al., 2020). Hydrogen has been applied in recycling and extracting of assembled NdFeB magnets from actual electronic devices, such as the hard disk drives and voice coil motors (Zakotnik et al., 2009; Gutfleisch et al., 2013).

Additive manufacturing has also attracted growing interests as a promising manufacturing method of NdFeB bulk magnets. Detailed information can be referred to the articles by Kwok et al. (2014), Jaćimović et al. (2017) and Volegov et al. (2020).

7.3 FABRICATION OF NdFeB SINTERED MAGNETS

There are two well-established approaches to manufacture commercial NdFeB magnets, including the powder metallurgy for anisotropic and full-dense sintered magnets, as well as the melt-spinning for isotropic bonded magnets. In Section 7.3 and 7.4, such two dominant fabrication techniques are discussed, with both fundamental and technological aspects considered. The preparation of NdFeB sintered magnets involves strip casting, hydrogen decrepitation, jet milling, alignment and compressing, sintering and post-sinter annealing.

7.3.1 Strip Casting

The $Nd_2Fe_{14}B$ matrix phase (T_1, with the atomic formula of $Nd_{11.76}Fe_{82.36}B_{5.88}$) is the ferromagnetic origin of the hard magnetism. Also, relative to the stoichiometric 2:14:1 composition, a certain surplus of Nd is required to generate the Nd-rich phase, acting as the liquid sintering aid to ensure rapid densification, and as the magnetic isolator of the neighboring ferromagnetic grains. Based on the NdFeB phase diagram with Nd:B = 2:1 (Figure 7.8), the solidification of $Nd_2Fe_{14}B$ phase enters the phase region of L + Fe, leading to the precipitation of the dendritic α-Fe phase in the ingot (Zhou and Dong, 2004). The α-Fe dendrites remain in the ferromagnetic state at the service temperature of NdFeB, which is detrimental to the coercivity. Presence of the ductile α-Fe also deteriorates the grindability of magnetic powders and hinders the grain alignment in the subsequent processing. Consequently, long-time homogenization annealing becomes necessary to remove the harmful α-Fe phase, which is however, costly and time-consuming.

To overcome the limitations of conventional ingots, strip casting (SC) technique has been developed, providing the following advantages. Firstly, fast cooling rate of the SC process (usually 10^3–10^6 °C/s) can suppress the formation of α-Fe dendrites and generates fine dispersion of Nd-rich phase along the columnar grains. As verified by the back-scattered SEM images in Figure 7.11, compared to the α-Fe dendrites with dark contrast and large RE-rich phase with bright contrast in the cast ingot, fine lamella microstructure where columnar grains with 5–25 μm width isolated by uniform RE-rich GBs are identified in the cast strips (Pei et al., 2002). Secondly, the SC process gives rise to directional crystalline growth and the resultant <00l> texture due to the temperature gradient from the wheel side to the free side. Overall, the SC technique enables the mass production of ingots with refined microstructure, and the preparation of close-to-stoichiometric NdFeB with high-ranking $(BH)_{max}$ over 400 kJ/m³ (Scott et al., 1996).

7.3.2 Hydrogen Decrepitation

The plate-like cast strips with typical thickness of 250–350 mm are crushed into friable powders of 10–100 μm by the hydrogen decrepitation (HD) technique. During the HD processing, the cast strips are placed in pure hydrogen atmosphere, where the Nd-rich intergranular phase hydrogenates initially, following the reaction of $Nd + (x/2)H_2 \rightarrow NdH_x + \Delta H$ at room temperature (Zhou and Dong, 2004). The temperature is then gradually increased to 150 °C, at which hydrogenation of the $Nd_2Fe_{14}B$ matrix phase starts, following $Nd_2Fe_{14}B + (x/2)H_2 \rightarrow Nd_2Fe_{14}BH_x + \Delta H$. The absorbed hydrogen as interstitial atoms in the lattice forms solid solution and increases the unit cell volume (Tanaka et al., 1997). Due to the different expansion rates of the two phases, the internal stress causes intergranular and transgranular fractures (Figure 7.10). The dispersion of Nd-rich phase along the GBs gives rise to fine particle size after HD processing, which is extremely brittle and can be readily milled afterward. Considering that the absorbed hydrogen lowers the H_A of the $Nd_2Fe_{14}B$ and the H_{cj} of the NdFeB sintered magnet, complete hydrogen desorption is required to recover the magnetic performance. Besides, the residual hydrogen that released during high-temperature sintering may also cause microcracks within the bulk magnets, exerting devastating effect on the mechanical properties. Consequently, the HD process usually involves both hydrogen absorption and desorption.

7.3.3 Jet Milling

The coarse particles after HD process are continuously fed into the fluidized-bed jet milling (JM) chamber for pulverization. Under the force of the high-velocity gas jets, the friable particles collide with each other. The milling vessel is integrated with a classifier, from which the particles with desired size are collected while the larger particles are returned for further milling. In comparison to ball milling, JM is more efficient and appropriate for scale-up production. Contaminants from the milling balls can also be largely reduced since the JM is mediated by the nitrogen or helium gas. The magnetic powders prepared by JM exhibits more uniform particle size distribution, typically around 3 µm in diameter. Owing to above advantages, the combination of SC, HD and JM techniques has been widely used in the industry of sintered NdFeB.

According to the nucleation theory, the coercivity is inversely proportional to $\ln D^2$ with D related to the average size of the 2:14:1 phase grains (Li et al., 2009). When the D is comparable to that of the single domain, the reversal mode of magnetic domains changes from domain wall motion to coherent rotation. Consequently, grain refinement has been generally acknowledged as an important strategy for coercivity enhancement. For the JM process, since the jet energy is inversely related to square root of the gas weight, replacing the traditional N_2 with lighter He yields higher jet energy and results in reduced particles sizes to ~1 µm. This gives rise to large coercivity of 2 T with the $(BH)_{max}$ approaching ~400 kJ/m^3 in the Dy-free NdFeB sintered magnets (Goto et al., 2010).

Note that very fine-grained particles are difficult to be aligned under magnetic field and are also easily oxidized during sintering, a critical diameter exists, below which the coercivity is abnormally decreased. This explains why the particle size is usually controlled at approximately 3 µm in mass production. On the laboratory scale, there is a growing trend toward ultrafine-grained (~1 µm) magnets produced by He JM, together with pressless process (PLP). By combining hydrogenation-disproportionation-desorption-recombination (HDDR), HD and He JM, the particle size can be further decreased to ~0.3 µm, being the regime of single domain size of the $Nd_2Fe_{14}B$ phase.

7.3.4 Alignment and Compressing

The large remanence B_r obtained by the power metallurgy procedure arises from the crystalline orientation of the 2:14:1 phase grains and high densification. Accordingly, the milled powders are consolidated into the cavity of a die mold to prepare the anisotropic green compact. Two common requirements have to be fulfilled during this process, one of which is a certain density to overcome the particle-particle friction via compressing under ~20 MPa. The other requirement is the grain orientation

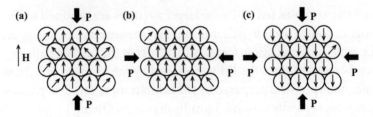

FIGURE 7.11 Schematic drawings showing various pressing routes, including (a) AP, (b) TP and (c) RIP.

(macroscopic easy axis) via magnetic alignment along the same direction of crystallographic c-axis under an applied magnetic field (~1.6 T). During the die pressing, the applied magnetic field can be either parallel (axial-field pressing, AP) or perpendicular (transverse-field pressing, TP) to the direction of compressing. For the AP approach (Figure 7.11a), the magnetic powders shift more easily during compacting, severely disturbing the grain alignment. Comparably, the TP approach has been used to compress relatively large rectangular blocks (Figure 7.11b). In 2000, rubber isostatic pressing (RIP) has been developed where the powders are sealed in a rubber mold under die press (Sagawa et al., 2000), and the uniaxial pressure is mediated through the rubber mold and converted to isostatic pressure (Figure 7.11c). This enables the production of near net-shape magnets with high texture degree comparable to isostatic pressing. For large samples however, cold isostatic pressing (CIP) is required, with applied pulsed magnetic field as high as 6–8 T. Through this way, maximum degree of alignment (99%) can be achieved with high remanence of 1.55 T (Kaneko, 2004), approaching the theoretical limit. In 2008, a pressless process has been proposed for fine powders (~1 μm) obtained by He JM (Sagawa and Une, 2008). The small grain size entitles the powders with high sintering activity to be consolidated without pressing. Such pressless processes are beneficial for reducing oxygen contamination, while its application is limited due to the high cost of generating fine powders and the large shrinkage rate during sintering.

7.3.5 Sintering

Liquid phase sintering is a vital consolidation process to fabricate full-dense bulk magnets, during which the green compacts are sintered in vacuum furnace. Upon increased temperature, the Nd-rich phase melts

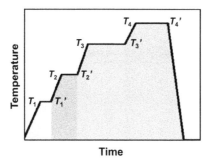

FIGURE 7.12 Design of the sintering curve for the NdFeB green compact.

into liquid, penetrating into the interface between adjacent grains under the capillary driving force, and the pores gradually diminish toward full densification. Considering the critical role of the liquid Nd-rich phase and its strong affinity with oxygen, the sintering atmosphere should be strictly controlled with pressure below 10^{-3} Pa. As shown in Figure 7.12, the whole sintering process includes several holding stages, i.e. between T_1–T_1' to remove the antioxidant additives (usually in the formula of C_nH_m), T_2–T_2' and T_3–T_3' to release the residual hydrogen within the $Nd_2Fe_{14}B$ matrix phase and Nd-rich intergranular phase, and T_4–T_4' for the designed sintering. Finally, the furnace is cooled down to room temperature under the protection of over-pressured Ar to minimize the atmospheric impurities such as O or N. The sintering temperature plays a crucial role on the microstructure and magnetic properties of ultimate magnet. Previous investigation has demonstrated that raised sintering temperature significantly affects the mobility of GBs, and alters the GB wetting of NdFeB from partial to complete state (Straumal et al., 2012). Consequently, continuous Nd-rich GB network forms to isolate neighboring ferromagnetic $Nd_2Fe_{14}B$ grains for enhanced coercivity. The sintering temperature is also influential to other microstructural features, including the magnet density, the $Nd_2Fe_{14}B$ grain size distribution, the structure of Nd-rich phase, which exert combined effects on the extrinsic B_r, H_{cj} and $(BH)_{max}$. Usually, the low-temperature sintering results in the low magnet density and agglomerated Nd-rich phase for low B_r, H_{cj} and $(BH)_{max}$ (Woodcock et al., 2012). High-temperature sintering above the optimal temperature leads to the exaggerated grain growth for reduced H_{cj} (Sepehri-Amin et al., 2011).

Differential scanning calorimetry (DSC) curve provides basic guideline to determine the sintering temperature. As typically shown in Figure 7.13,

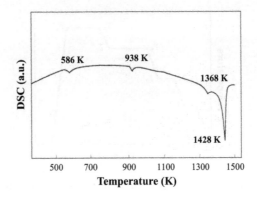

FIGURE 7.13 Typical DSC curve of the $Nd_{15}Fe_{78}B_7$ alloy. Reprinted with permission from (Kaneko and Ishigaki, 1994). Copyright (1994) Springer Nature.

four endothermic peaks can be observed (Kaneko and Ishigaki, 1994). The first peak at 586 K corresponds to the Curie transition of the 2:14:1 phase from ferromagnetic to paramagnetic state. The second peak at 938 K relates to the ternary eutectic temperature (L → 2:14:1 phase + B-rich phase + Nd-rich phase). The third peak at 1368 K represents the binary eutectic temperature (L → 2:14:1 phase + B-rich phase). The fourth peak at 1428 K corresponds to the melting temperature of the 2:14:1 phase. Consequently, the sintering temperature should be chosen between the temperate interval of the third and fourth peaks, specifically, below the melting point of $RE_2Fe_{14}B$ matrix phase to maintain the tetragonal crystal structure.

7.3.6 Post-Sinter Annealing

A two-step post-sinter annealing (PSA) treatment has been proved to effectively enhance the coercivity of NdFeB (Jin et al., 2019b). For example, the Dy-free sample containing 0.1 at% Ga exhibits increased coercivity from 1.5 to 2.0 T after annealing at 1093 K for 3 h and 763 K for 4 h (Xu et al., 2018). Previous studies have suggested that the first PSA above the ternary eutectic temperature of ~938 K yields the formation of minor liquid phase, and generates continuous GB phase of several nanometers (Shinba et al., 2005). The second PSA below the ternary eutectic point contributes to smoothing the GBs and decreasing the defect density at the interface of the $Nd_2Fe_{14}B$/Nd-rich phases, hence suppressing the nucleation of reversal domains. Considering the critical role of Nd-rich phase for coercivity

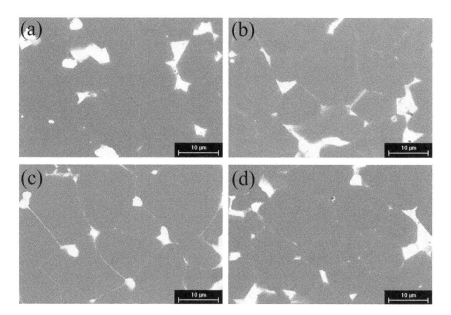

FIGURE 7.14 Back-scattered SEM images of Al/Cu-co-doped magnet, (a) before annealing, and after annealing at (b) 450 °C, (c) 480 °C, and (d) 550 °C. Reprinted with permission from (Ni et al., 2010). Copyright (2010) Elsevier.

enhancement, the evolution of different compositions and crystal structures of Nd-rich phase during two-step PSA has been extensively studied, including the metallic Nd with double hexagonal close packed (dhcp) or face-centered cubic (fcc) structures, a-type Nd_2O_3 (hP5), c-type Nd_2O_3 (cI80) and NdO (cF8) (Mo et al., 2008). Kim et al. have demonstrated the phase transformation of Nd-rich triple junction phase from hexagonal to cubic Nd_2O_3 phase after PSA, which reduces the lattice misfit and the mechanical stress at the interface (Kim et al., 2012; Hrkac et al., 2014). With the assistance of scanning TEM and three-dimensional atom probe, minor Cu additive has been found to be segregated at the GBs after PSA, which is expected to stabilize the different crystal structures (Sepehri-Amin et al., 2012). Note that the optimal PSA is heavily dependent on the composition design, particularly for the Cu/Ga-doped magnets. Figure 7.14 illustrates typical SEM images taken from the cross-sectional Al/Cu-co-doped magnets after annealing at different temperatures, and the optimal annealing at 753 K leads to the penetration of RE-rich phase into the $RE_2Fe_{14}B$ interfaces (Ni et al., 2010).

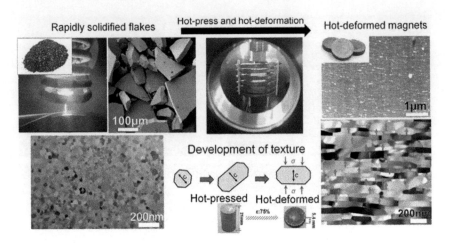

FIGURE 7.15 Schematic production process of the NdFeB hot-pressed and hot-deformed magnets. Reprinted with permission from (Sepehri-Amin et al., 2018). Copyright (2018) Elsevier.

7.3.7 Commercial NdFeB Sintered Magnets

The grade of commercial NdFeB sintered magnets includes N50, 48M, 45H, 42SH, 42UH, 40EH, 38TH etc., with the numerical values corresponding to the maximum energy product (in the CGS unit of MGOe), and the capital letters representing the maximum service temperatures with N = 80 °C, M = 100 °C, H = 120 °C, SH = 150 °C, UH = 180 °C, EH = 200 °C, and TH = 220 °C.

7.4 FABRICATION OF NdFeB BONDED MAGNETS

As a practical alternative to the powder metallurgy technique, melt-spinning is also commercially utilized to produce rapidly solidified ribbons with metastable nanostructure or amorphous state. The molten alloy is ejected onto the rotating water-cooled copper wheel, utilizing the conductive nature of the Cu wheel to extract heat from the ribbon. The resultant microstructure is highly dependent on the cooling rates, spanning from coarsely crystalline at low speed (under-quenched) to nanocrystalline at intermediate speed (optimally quenched) and amorphous at high speed (over-quenched). For the under-quenched state, the microstructure varies from nanocrystalline (wheel side) to coarsely crystalline (free side) by 2 ~ 3 orders of magnitude. For the optimally quenched state, most grains

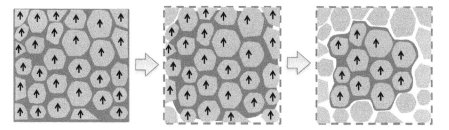

FIGURE 7.16 Schematics showing the corrosion process of NdFeB sintered magnets, with the low-electrode-potential Nd-rich phase preferentially corroded in corrosive environments, followed by the detachment of loosened 2:14:1 matrix phase grains.

exhibit the near-optimal grain size (typically ~20 nm) to provide the high coercivity. For the over-quenched state, substantial undercooling of the melt induces an amorphous structure with magnetically soft nature, followed by short-time crystallization annealing. When the appropriate quenching speed is chosen, the required single-domain nanostructure can be obtained for higher coercivity.

The melt-spun ribbons with isotropic nature are suitable to mix with polymers to fabricate the bonded NdFeB magnets with medium magnetic properties, i.e. the ribbon fragments are glued together with dry epoxy resin followed by cold-compressing. If the rapidly solidified flakes are hot-pressed, followed by hot-deformation, anisotropic nanocrystalline magnets with c-axis grain alignment parallel to the pressing direction can be achieved (Figure 7.15). During the hot-pressing, the ribbon fragments are loaded into the die, heated to desired temperature and pressed under uniaxial pressure. Due to the advantageous single-domain nanostructure, it is anticipated that the melt-spinning technique should gain increasing importance in the near future to produce the bulk magnets.

7.5 SURFACE COATING

Poor corrosion resistance is one of the major drawbacks of the NdFeB magnets. Corrosion behavior of the NdFeB sintered magnet in different environments such as in neutral, acid and alkaline solutions have been studied in the literature (Schultz et al., 1999). The lower electrode potential of Nd-rich intergranular phase than that of the $Nd_2Fe_{14}B$ matrix phase provides the driving force for the electrochemical corrosion of NdFeB magnets (Yan et al., 2009). Figure 7.16 shows the corrosion behavior of conventional NdFeB, where the chemically active Nd-rich phase is preferentially corroded, followed by the detachment of the loosened $Nd_2Fe_{14}B$ matrix phase grains from the bulk NdFeB, causing the eventual disintegration of the magnet. Consequently, in practical applications, particularly for the offshore wind turbine working in corrosive environments, surface coating is necessary. Before coating, machining is required to shape the final magnet with designated geometry. Different types of surface coatings depending on the application conditions have been developed, including the Zn, Ni, Ni-Cu-Ni and epoxy resins. Generally, the metallic coatings provide better mechanical stability, while the polymeric coatings are more cost efficient.

REFERENCES

Chen, F. G., T. Q. Zhang, W. H. Zhang, L. T. Zhang, and Y. X. Jin. 2018. Dependence of the demagnetization behavior on the direction of grain boundary diffusion in sintered Nd-Fe-B magnets. *Journal of Magnetism and Magnetic Materials* 465:392–8.

Fan, X. D., G. F. Ding, K. Chen, S. Guo, C. Y. You, R. J. Chen, D. Lee, and A. R. Yan. 2018. Whole process metallurgical behavior of the high-abundance rare-earth elements LRE (La, Ce and Y) and the magnetic performance of $Nd_{0.75}LRE_{0.25}$-Fe-B sintered magnets. *Acta Materialia* 154:343–54.

Fu, S., X. L. Liu, J. Y. Jin, Z. H. Zhang, Y. S. Liu, and M. Yan. 2021. Magnetic properties evolution with grain boundary phase transformation and their growth in Nd-Fe-Cu-Ga-B sintered magnet during post-sinter annealing process. *Intermetallics* 137:107303.

Fukuno, A., K. Hirose, and T. Yoneyama. 1990. Coercivity mechanism of sintered NdFeB magnets having high coercivities. *Journal of Applied Physics* 67(9):4750–2.

Goto, S., S. Matsuura, N. Tezuka, Y. Une, and M. Sagawa. 2010. Microstructure during the fabrication process of Nd-Fe-B sintered magnets using atomized powders, in *21st Workshop on Rare-Earth Permanent Magnets and their Applications*, Slovenia.

Hallemans, B., P. Wollants, and J. R. Roos. 1995. Thermodynamic assessment of the Fe-Nd-B phase diagram. *Journal of Phase Equilibria* 16(2):137–49.

Harris, I. R., and G. W. Jewell. 2012. Rare-earth magnets: properties, processing and applications. In *Functional materials for sustainable energy applications*, eds. J. A. Kilner, S. J. Skinner, S. J. C. Irvine, and P. P. Edwards, 600–39. Cambridge: Woodhead Publishing.

Herbst, J. F. 1991. $R_2Fe_{14}B$ materials: Intrinsic properties and technological aspects. *Review of Modern Physics* 63(4):819–98.

Hirota, K., H. Nakamura, T. Minowa, and M. Honshima. 2006. Coercivity enhancement by the grain boundary diffusion process to Nd-Fe-B sintered magnets. *IEEE International Magnetics Conference* 42:2909–11.

Hrkac, G., T. G. Woodcock, K. T. Butler, L. Saharan, M. T. Bryan, T. Schrefl, and O. Gutfleisch. 2014. Impact of different Nd-Rich crystal-phases on the coercivity of Nd-Fe-B grain ensembles. *Scripta Materialia* 70:35–8.

Hu, J. F., Y. L. Liu, M. L. Yin, Y. Z. Wang, B. P. Hu, and Z. X. Wang. 1999. Investigation of simultaneous enhancement of remanence and coercivity in Nd-Fe-B/Nb mixed-powder sintered magnets. *Journal of Alloys and Compounds* 288(1–2):226–8.

Jaćimović, J., F. Binda, L. G. Herrmann, F. Greuter, J. Genta, M. Calvo, T. Tomše, and R. A. Simon. 2017. Net shape 3D printed NdFeB permanent magnet. *Advanced Engineering Materials* 19(8):1700098.

Jin, J. Y., M. Yan, T. Y. Ma, W. Li, Y. S. Liu, Z. H. Zhang, and S. Fu. 2020. Balancing the microstructure and chemical heterogeneity of multi-main-phase Nd-Ce-La-Fe-B sintered magnets by tailoring the liquid-phase-sintering, *Materials and Design* 186:108308.

Jin, J. Y., M. Yan, W. Chen, W. Y. Zhang, Z. H. Zhang, L. Z. Zhao, G. H. Bai, and J. M. Greneche. 2021. Grain boundary engineering towards high-figure-of-merit Nd-Ce-Fe-B sintered magnets: Synergetic effects of (Nd, Pr)H$_x$ and Cu co-dopants. *Acta Materialia* 204:116529.

Jin, J. Y., M. Yan, Y. S. Liu, B. X. Peng, and G. H. Bai. 2019a. Attaining high magnetic performance in as-sintered multi-main-phase Nd-La-Ce-Fe-B magnets: Toward skipping the post-sinter annealing treatment. *Acta Materialia* 169:248–59.

Jin, J. Y., T. Y. Ma, Y. J. Zhang, G. H. Bai, and M. Yan. 2016a. Chemically inhomogeneous RE-Fe-B permanent magnets with high figure of merit: solution to global rare earth criticality. *Scientific Reports* 6:32200.

Jin, J. Y., Y. J. Zhang, G. H. Bai, Z. Y. Qian, C. Wu, T. Y. Ma, B. G. Shen, and M. Yan. 2016b. Manipulating Ce valence in RE$_2$Fe$_{14}$B tetragonal compounds by La-Ce co-doping: resultant crystallographic and magnetic anomaly. *Scientific Reports* 6:30194.

Jin, J. Y., Y. M. Tao, X. H. Wang, Z. Y. Qian, W. Chen, C. Wu, and M. Yan. 2022. Concurrent improvements of corrosion resistance and coercivity in Nd-Ce-Fe-B sintered magnets through engineering the intergranular phase. *Journal of Materials Science & Technology* 110:239–45.

Jin, J. Y., Z. H. Zhang, L. Z. Zhao, B. X. Peng, Y. S. Liu, J. M. Greneche, and M. Yan. 2019b. Evolution of REFe$_2$ (RE=rare earth) phase in Nd-Ce-Fe-B magnets and resultant Ce segregation. *Scripta Materialia* 170:150–5.

Kaneko, Y. 2004. Technological evolution and application trends of NdFeB sintered magnets in Japan, in *Proceedings of the 18th Workshop on High Performance Magnets & Their Applications*, France.

Kaneko, Y., and N. Ishigaki. 1994. Recent developments of high-performance NEOMAX magnets. *Journal of Materials Engineering and Performance* 3(2):228–33.

Kobayashi, K., K. Urushibata, Y. Une, and M. Sagawa. 2013. The origin of coercivity enhancement in newly prepared high coercivity Dy-free Nd-Fe-B sintered magnets. *Journal of Applied Physics* 113:163910.

Kim, S. H., J. W. Kim, T. S. Jo, and Y. D. Kim. 2011. High coercive Nd-Fe-B magnets fabricated via two step sintering. *Journal of Magnetism and Magnetic Materials* 323(22):2851–4.

Kim, T. H., S. R. Lee, S. Namkumg, and T. S. Jang. 2012. A study on the Nd-rich phase evolution in the Nd-Fe-B sintered magnet and its mechanism during post-sintering annealing. *Journal of Alloys and Compounds* 537:261–8.

Kim, T. H., T. T. Sasaki, T. Koyama, Y. Fujikawa, and K. Hono. 2019. Formation mechanism of Tb-rich shell in grain boundary diffusion processed Nd-Fe-B sintered magnets. *Scripta Materialia* 178:433–7.

Kwok, S. W., S. A. Morin, B. Mosadegh, J. So, R. F. Shepherd, R. V. Martinez, B. Smith, F. C. Simeone, A. A. Stokes, and G. M. Whitesides. 2014. Magnetic assembly of soft robots with hard components. *Advanced Functional Materials* 24:2180–7.

Li, J., X. Tang, H. Sepehri-Amin, T. T. Sasaki, T. Ohkubo, and K. Hono. 2020. Angular dependence and thermal stability of coercivity of Nd-rich Ga-doped Nd-Fe-B sintered magnet. *Acta Materialia* 187:66–72.

Liang, L. P., T. Y. Ma, P. Zhang, and M. Yan. 2015. Effects of $Dy_{71.5}Fe_{28.5}$ intergranular addition on the microstructure and the corrosion resistance of Nd-Fe-B sintered magnets. *Journal of Magnetism and Magnetic Materials* 384:133–7.

Liang, L. P., T. Y. Ma, P. Zhang, J. Y. Jin, and M. Yan. 2014. Coercivity enhancement of NdFeB sintered magnets by low melting point $Dy_{32.5}Fe_{62}Cu_{5.5}$ alloy modification. *Journal of Magnetism and Magnetic Materials* 355:131–5.

Liu, P., T. Y. Ma, X. H. Wang, Y. J. Zhang, and M. Yan. 2015. Role of hydrogen in Nd-Fe-B sintered magnets with DyH_x addition. *Journal of Alloys and Compounds* 628:282–6.

Liu, X. L., X. J. Wang, L. P. Liang, P. Zhang, J. Y. Jin, Y. J. Zhang, T. Y. Ma, and M. Yan. 2014. Rapid coercivity increment of Nd-Fe-B sintered magnets by $Dy_{69}Ni_{31}$ grain boundary restructuring. *Journal of Magnetism and Magnetic Materials* 370:76–80.

Liu, X. L., X. W. Wu, J. Y. Jin, Y. M. Tao, X. H. Wang, and M. Yan. 2022. Microstructure and magnetic performance of Nd-Y-Ce-Fe-B sintered magnets after annealing. *Rare Metals* 41(3):859–64.

Liu, Y. K., X. F. Liao, J. Y. He, H. Y. Yu, X. C. Zhong, Q. Zhou, and Z. W. Liu. 2020a. Magnetic properties and microstructure evolution of in-situ Tb-Cu diffusion treated hot-deformed Nd-Fe-B magnets. *Journal of Magnetism and Magnetic Materials* 504:166685.

Liu, Y. S., J. Y. Jin, M. Yan, M. X. Li, B. X. Peng, Z. H. Zhang, and X. H. Wang. 2020b. A reliable route for relieving the constraints of multi-main-phase Nd-La-Ce-Fe-B sintered magnets at high La-Ce substitution: (Pr, Nd)H_x grain boundary diffusion. *Scripta Materialia* 185:122–8.

Liu, Y. S., J. Y. Jin, T. Y. Ma, B. X. Peng, X. H. Wang, and M. Yan. 2021. Promoting the La solution in 2:14:1-type compound: resultant chemical deviation and microstructural nanoheterogeneity. *Journal of Materials Science & Technology* 62:195–202.

Li, W. F., T. Ohkubo, K. Hono, and M. Sagawa. 2009. The origin of coercivity decrease in fine grained Nd-Fe-B sintered magnets. *Journal of Magnetism and Magnetic Materials* 321(8):1100–5.

Li, W. F., H. Sepehri-Amin, T. Ohkubo, N. Hase, and K. Hono. 2011. Distribution of Dy in high-coercivity (Nd,Dy)-Fe-B sintered magnet. *Acta Materialia* 59:3061–9.

Lu, K. C., X. Q. Bao, X. F. Song, Y. X. Wang, M. H. Tang, J. H. Li, and X. X. Gao. 2021. Temperature dependences of interface reactions and Tb diffusion behavior of Pr-Tb-Cu-Al alloys/Nd-Fe-B magnet. *Scripta Materialia* 191:90–5.

Ma, T. Y., X. J. Wang, X. L. Liu, C. Wu, and M. Yan. 2015. Coercivity enhancements of Nd-Fe-B sintered magnets by diffusing DyH_x along different axes. *Journal of Physics D: Applied Physics* 48:215001.

Mo, W. J., L. T. Zhang, Q. Z. Liu, A. D. Shan, J. S. Wu, and M. Komuro. 2008. Dependence of the crystal structure of the Nd-rich phase on oxygen content in an Nd-Fe-B sintered magnet. *Scripta Materialia* 59(2):179–82.

Ni, J. J., T. Y. Ma, and M. Yan. 2012. Improvement of corrosion resistance in Nd-Fe-B magnets through grain boundaries restructuring. *Materials Letters* 75:1–3.

Ni, J. J., T. Y. Ma, Y. R. Wu, and M. Yan. 2010. Effect of post-sintering annealing on microstructure and coercivity of $Al_{85}Cu_{15}$-added Nd-Fe-B sintered magnets. *Journal of Magnetism and Magnetic Materials* 322(22):3710–3.

Gutfleisch, O., K. Güth, T. G. Woodcock, and L. Schultz. 2013. Recycling used Nd-Fe-B sintered magnets via a hydrogen-based route to produce anisotropic, resin bonded magnets. *Advanced Energy Materials* 3:151–5.

Okada, M., S. Sugimoto, C. Ishizaka, T. Tanaka, and M. Homma. 1985. Didymium-Fe-B sintered permanent magnets. *Journal of Applied Physics* 57(1):4146–8.

Oono, N., M. Sagawa, R. Kasada, H. Matsui, and A. Kimura. 2011. Production of thick high performance sintered neodymium magnets by grain boundary diffusion treatment with dysprosium-nickel-aluminum alloy. *Journal of Magnetism and Magnetic Materials* 323(3–4):297–300.

Pathak, A. K., M. Khan, K. A. Gschneidner Jr, R. W. McCallum, L. Zhou, K. Sun, K. W. Dennis, C. Zhou, F. E. Pinkerton, M. J. Kramer, and K. Pecharsky. 2015. Cerium: an unlikely replacement of Dysprosium in high performance Nd-Fe-B permanent magnets. *Advanced Materials* 27:2663–7.

Pei, W. L. C. S. He, F. Z. Lian, G. Q., G. Zhou, and H. C. Yang. 2002. Structures and magnetic properties of sintered Nd-Fe-B magnets produced by strip casting technique. *Journal of Magnetism and Magnetic Materials* 239(1–3):475–8.

Peng, B. X., J. Y. Jin, Y. S. Liu, C. X. Lu, L. W. Li, and M. Yan. 2020. Towards peculiar corrosion behavior of multi-main-phase Nd-Ce-Y-Fe-B permanent material with heterogeneous microstructure. *Corrosion Science* 177:108972.

Peng, B. X., T. Y. Ma, Y. J. Zhang, J. Y. Jin, and M. Yan. 2017. Improved thermal stability of Nd-Ce-Fe-B sintered magnets by Y substitution. *Scripta Materialia* 131:11–4.

Ramesh, R., and K. Srikrishna. 1988. Magnetization reversal in nucleation controlled magnets. I. Theory. *Journal of Applied Physics* 64(11):6406–15.

Sagawa, M., H. Naagta, T. Watanabe, and O. Itatani. 2000. Rubber isostatic pressing (RIP) of powders for magnets and other materials. *Materials and Design* 21:243–9.

Sagawa, M., and Y. Une. 2008. A new process for producing Nd-Fe-B sintered magnets with small grain size, in *20th International Workshop on Rare Earth Permanent Magnets and their Applications*, Greece.

Sasaki, T. T., T. Ohkubo, Y. Takada, T. Sato, A. Kato, Y. Kaneko, and K. Hono. 2016. Formation of non-ferromagnetic grain boundary phase in a Ga-doped Nd-rich Nd-Fe-B sintered magnet. *Scripta Materialia* 113:218–21.

Schultz, L., A. M. El-Aziz, G. Barkleit, and K. Mummert. 1999. Corrosion behavior of Nd-Fe-B permanent magnetic alloys. *Materials Science & Engineering A* 267(2):307–13.

Scott, D. W., B. M. Ma, Y. L. Liang, and C. O. Bounds. 1996. Microstructural control of NdFeB cast ingots for achieving 50 MGOe sintered magnets. *Journal of Applied Physics* 79:4830.

Sepehri-Amin, H., Y. Une, T. Ohkubo, K. Hono, and M. Sagawa. 2011. Microstructure of fine-grained Nd-Fe-B sintered magnets with high coercivity. *Scripta Materialia* 65(5):396–9.

Sepehri-Amin, H., T. Ohkubo, T. Shima, and K. Hono. 2012. Grain boundary and interface chemistry of an Nd-Fe-B-based sintered magnet. *Acta Materialia* 60:819–30.

Sepehri-Amin, H., T. Ohkubo, and K. Hono. 2013. The mechanism of coercivity enhancement by the grain boundary diffusion process of Nd-Fe-B sintered magnets. *Acta Materialia* 61(6):1982–90.

Shinba, Y., T. J. Konno, K. Ishikawa, K. Hiraga, and M. Sagawa. 2005. Transmission electron microscopy study on Nd-rich phase and grain boundary structure of Nd-Fe-B sintered magnets. *Journal of Applied Physics* 97(5):053504.

Sodernik, K. A., K. U. Roman, M. Komelj, A. Kovács, and S. Turm. 2021. Microstructural insights into the coercivity enhancement of grain-boundary-diffusion-processed Tb-treated Nd-Fe-B sintered magnets beyond the core-shell formation mechanism. *Journal of Alloys and Compounds* 864:158915.

Straumal, B. B., Y. O. Kucheev, I. L. Yatskovskaya, I. V. Mogilnikova, G. Schutz, A. N. Nekrasov, and B. Baretzky. 2012. Grain boundary wetting in the NdFeB-based hard magnetic alloys. *Journal of Materials Science* 47:8352–9.

Sugimoto, S. 2011. Current status and recent topics of rare-earth permanent magnets. *Journal of Physics D: Applied Physics* 44(6):064001.

Tanaka, K., Y. Hayashi, M. Kimura, and M. Yamada. 1997. Hydrogen absorbing and desorbing properties of Nd-Fe-B and Nd-Co-B amorphous alloys. *Journal of Alloys and Compounds* 253–254:101–5.

Tang, W., S. Zhou, and R. Wang. 1989. Preparation and microstructure of La-containing R-Fe-B permanent magnets. *Journal of Applied Physics* 65(8):3142–5.

Volegov, A. S., S. V. Andreeva, N. V. Seleznevaa, I. A. Ryzhikhina, N. V. Kudrevatykha, L. Madler, I. V. Okulov. 2020. Additive manufacturing of heavy rare earth free high-coercivity permanent magnets. *Acta Materialia* 188:733–9.

Woodcock, T. G., Y. Zhang, G. Hrkac, G. Ciuta, N. M. Dempsey, T. Schrefl, O. Gutfleisch, and D. Givord. 2012. Understanding the microstructure and coercivity of high performance NdFeB-based magnets. *Scripta Materialia* 67:536–41.

Wu, X. W., J. Y. Jin, Y. M. Tao, W. Chen, X. L. Peng, and M. Yan. 2021. High synergy of coercivity and thermal stability in resource-saving Nd-Ce-Y-Fe-B melt-spun ribbons. *Journal of Alloys and Compounds* 882:160731.

Xu, X. D., T. T. Sasaki, J. N. Li, Z. J. Dong, H. Sepehri-Amin, T. H. Kim, T. Ohkubo, T. Schrefl, and K. Hono. 2018. Microstructure of a Dy-free Nd-Fe-B sintered magnet with 2 T coercivity. *Acta Materialia* 156:146–57.

Yan, G. L., P. J. McGuiness, J. P. G. Farr, and I. R. Harris. 2009. Environmental degradation of NdFeB magnets. *Journal of Alloys and Compounds* 478(1–2): 188–92.

Yan, M., J. Y. Jin, and T. Y. Ma. 2019. Grain boundary restructuring and La/Ce/Y application in Nd-Fe-B magnets. *Chinese Physics B* 28(7):077507.

Yin, X. W., M. Yue, Q. M. Lu, M. Liu, F. Wang, Y. B. Qiu, W. Q. Liu, T. Y. Zuo, S. S. Zha, X. L. Li, and X. F. Yi. 2020. An efficient process for recycling Nd-Fe-B sludge as high-performance sintered magnets. *Engineering* 6:165–72.

Zakotnik, M., I. R. Harris, and A. J. Williams. 2009. Multiple recycling of NdFeB-type sintered magnets. *Journal of Alloys and Compounds* 469(1–2):314–21.

Zhao, Y., H. B. Feng, A. H. Li, and W. Li. 2020. Microstructural and magnetic property evolutions with diffusion time in TbH_x diffusion processed Nd-Fe-B sintered magnets. *Journal of Magnetism and Magnetic Materials* 515:167272.

Zhang, Y. J., T. Y. Ma, M. Yan, J. Y. Jin, B. Wu, B. X. Peng, Y. S. Liu, M. Yue, and C. Y, Liu. 2018. Post-sinter annealing influences on coercivity of multi-main-phase Nd-Ce-Fe-B magnets. *Acta Materialia* 146:97–105.

Zhang, Z. H., J. Y. Jin, L. P. Liang, B. X. Peng, Y. S. Liu, S. Fu, and M. Yan. 2019. High-performance Nd-Fe-B sintered magnets via co-doping high-melting-point Zr and low-melting-point $Dy_{71.5}Fe_{28.5}$. *Journal of Magnetism and Magnetic Materials* 487:165356.

Zhang, Z. H., J. Y. Jin, T. Y. Ma, L. P. Liang, and M. Yan. 2020. Nd-Fe-B sintered magnets with low rare earth content fabricated via $Dy_{71.5}Fe_{28.5}$ grain boundary restructuring. *Journal of Magnetism and Magnetic Materials* 498:166162.

Zhou, S. Z., and Q. F. Dong. 2004. *Super permanent magnet: rare earth-iron based permanent magnetic material*. Beijing: Metallurgical Industry Press.

Zhu, M., W. Li, J. Wang, L. Zheng, Y. Li, K. Zhang, H. Feng, and T. Liu. 2014. Influence of Ce content on the rectangularity of demagnetization curves and magnetic properties of Re-Fe-B magnets sintered by double main phase alloy method. *IEEE Transactions on Magnetic* 50(1):1000104.

FURTHER READING

Sepehri-Amin, H., Hirosawa, S., and Hono, K. 2018. Advances in Nd-Fe-B based permanent magnets. In *Handbook of magnetic materials*, ed. E. Brück, 27:269–372. Amsterdam: North-Holland Publishing.

CHAPTER **8**

Other Emerging Hard Magnetic Materials

8.1 NANOCOMPOSITES

8.1.1 Introduction

Over the past two decades, the exponential growth of $(BH)_{max}$ has stalled and there seems to be little scope for any further dramatic $(BH)_{max}$ improvement for bulk permanent magnets such as NdFeB, which pushes the development of oriented hard/soft nanocomposites. By fully exploiting the nanoscale exchange coupling effect between hard phase with high magnetocrystalline anisotropy H_A and soft phase with high saturation magnetization M_s, novel nanocomposite hard magnetic materials with record-high $(BH)_{max}$ are anticipated. In this section, the milestones in both theoretical calculations and practical experiments of nanocomposites will be reviewed, from fundamental research, synthesis technique to current challenges.

8.1.2 Timeline of Nanocomposites Development

In the late 1980s, Coehoorn et al. prepared the first isotropic $Nd_2Fe_{14}B$/Fe_3B nanocomposite magnets with decent M_r/M_s ratio of 0.7–0.8, exceeding that (0.5) for the randomly oriented non-interacting crystallites (Coehoorn et al., 1988; Buschow et al., 1988; Coehoorn et al., 1989). Such significant remanence enhancement enlightens the strong interactions between the crystallites. A one-dimensional model has

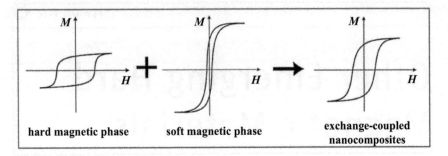

FIGURE 8.1 Combination of the hard and soft magnetic phases into the exchange-coupled nanocomposites. Reprinted with permission from (Liu et al., 2014). Copyright (2014) Royal Society of Chemistry.

been proposed to explain the nanoscale exchange coupling through the interfaces of hard and soft magnetic phases (Kneller and Hawig, 1991). On condition that a soft magnetic phase with high M_s is thin enough, its magnetic moment rotates coherently along the magnetization direction of neighboring hard magnetic phase with high H_A (Bader, 2006). The nanocomposite behaves like a single-phase permanent magnet toward enlarged $(BH)_{max}$, and no occurrence of kinks can be identified from the hysteresis loop, as typically displayed in Figure 8.1 (Liu et al., 2014). From the results of micromagnetic simulation, the size of soft magnetic phase should be smaller than twice the domain width of hard magnetic phase, meanwhile the soft and hard magnetic phases should be uniformly distributed in the design of the nanocomposite (Schrefl et al., 1994; Fischer et al., 1998).

Numerous calculations have been conducted to predict the theoretical $(BH)_{max}$ of nanocomposites based on exchange coupling of different hard and soft magnetic phases. In 1993, a three-dimensional model of anisotropic exchange coupling predicted the theoretical $(BH)_{max}$ of 137 MGOe for the oriented $Sm_2Fe_{17}N_3/Fe_{65}Co_{35}$ nanocomposite magnet (Skomski and Coey, 1993b; 1993c). In 1998, the theoretical $(BH)_{max}$ of the anisotropic FePt/Fe and SmCo/FeCo layered systems have been calculated to be 90 and 65 MGOe, respectively (Sabiryanov and Jaswal, 1998a; 1998b). In 2006, the theoretical $(BH)_{max}$ of anisotropic $Pr_2Fe_{14}B/\alpha$-Fe and $Sm_2Fe_{17}N_3/Fe_{65}Co_{35}$ have been calculated to be 101 and 112 MGOe, respectively (Rong et al., 2006). For the $SmCo_5/\alpha$-Fe core-shell structured nanocomposite magnet, the theoretical $(BH)_{max}$ has been simulated as ~100 MGOe at 300 K and ~88 MGOe at 473 K, which is prospective for high-temperature applications (Fukunaga et al., 2013). Such predictions demonstrate that the theoretical $(BH)_{max}$ of nanocomposite magnets far exceeds their corresponding single-phase magnets, providing new opportunities for the future development of super-strong permanent materials.

8.1.3 Synthesis Techniques

Accompanied with the theoretical calculations, experimental studies of nanocomposites through melt-spinning, mechanical alloying, thin film sputtering and chemical synthesis are ongoing, mainly involving the (Nd/Pr)$_2$Fe$_{14}$B/α-Fe, (Nd/Pr)$_2$Fe$_{14}$B/Fe$_3$B, Sm$_2$Fe$_{17}$N$_x$/α-Fe and FePt/Fe$_3$Pt nanocomposites.

In the early research, isotropic Nd$_2$Fe$_{14}$B/α-Fe nanocomposite ribbon was prepared by melt-spinning, with $(BH)_{max}$ of 20 MGOe and H_{cj} of 0.61 T (Manaf et al., 1993). The two-phase microstructure with ultrafine α-Fe (< 10 nm) distributed at the grain boundaries of Nd$_2$Fe$_{14}$B matrix phase (< 30 nm) facilitates the strong exchange coupling for single-phase hysteresis loop. Isotropic Sm$_2$Fe$_{17}$N$_x$/α-Fe magnet with $(BH)_{max}$ of 26 MGOe has also been prepared from the mechanically alloyed SmFe powders, followed by heat treatment and nitridation (Ding et al., 1993).

Different from the above top-down techniques, bottom-up chemical synthesis method involves initial nanoparticles, followed by nanocrystalline magnets through self-assembly, compaction or sintering (Sun et al., 2000), which gives rise to small particles (down to 2 nm) with narrow size distribution and satisfactory crystallinity (Liu, 2009). In 2002, isotropic FePt/Fe$_3$Pt nanocomposite has been chemically synthesized, via the self-assembly of 4 nm-Fe$_{58}$Pt$_{42}$ and 4 nm-Fe$_3$O$_4$ nanoparticles, and subsequent annealing at 650 °C in reduction atmosphere (95% Ar + 5% H$_2$) (Zeng et al., 2002). The controllable dispersion of soft Fe$_3$Pt in the hard FePt matrix phase gives rise to the $(BH)_{max}$ of 20.1 MGOe, which is over 50% higher than the theoretical limit of the non-exchange-coupled FePt counterpart. In 2016, isotropic nanocomposite consisting of hard magnetic Nd$_2$Fe$_{14}$B core of ~45 nm and soft magnetic α-Fe shell of ~13 nm exhibits a record-high $(BH)_{max}$ of 25 MGOe (Li et al., 2016).

The research on nanocomposite multilayer magnets have met with considerable success, due to the merits of tunable composition and microstructure by designing different magnetic layers with adjustable thicknesses, as well as selecting different heat treatments. The (Nd, Dy)(Fe, Co, Nb, B)$_{5.5}$/α-Fe nanocomposite film has been fabricated by sputtering deposition and heat treatment, achieving high B_r of 1.31 T, H_{cj} of 1.85

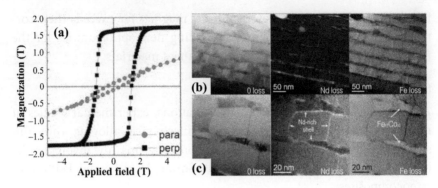

FIGURE 8.2 (a) Hysteresis loops of multilayer films Ta/Nd$_2$Fe$_{14}$B/Nd/Ta/Fe$_{67}$Co$_{33}$/Ta/NdFeB/Nd/Ta, (b and c) corresponding cross-sectional TEM images. Reprinted with permission from (Cui et al., 2012). Copyright (2012) John Wiley and Sons.

T, and $(BH)_{max}$ of 25 MGOe (Liu et al., 2002). The fully epitaxially grown SmCo$_5$/Fe multilayer films have been fabricated by ultra-high voltage pulsed laser deposition with $(BH)_{max}$ up to 50 MGOe, which is much higher than that of the theoretical value (31 MGOe) of anisotropic sing-phase SmCo$_5$ magnet (Neu et al., 2012). Later on, the Ta interlayer has been introduced to form the Ta(50 nm)/[NdFeB(30 nm)/Nd(3 nm)/Ta(1 nm)/Fe$_2$Co(10 nm)/Ta(1 nm)]$_N$/NdFeB(30 nm)/Nd(3 nm)/Ta(20 nm) (N = 9) multilayer films with $(BH)_{max}$ of 61 MGOe Figure 8.2a by magnetron sputtering, which is the highest $(BH)_{max}$ for nanocomposites currently (Cui et al., 2012). Microstructural characterizations in Figures 8.2b and c reveal that thin Nd-rich grain boundary layer forms to cover the Nd$_2$Fe$_{14}$B phase. Ascribing to the inserted Ta spacer layer, the Nd-rich intergranular phase is continuous and stable during grain boundary diffusion process at ~650 ° C, without alloying with the soft Fe$_2$Co layer.

8.1.4 Challenges and Perspectives

Great progress has been achieved in preparing low-dimensional hard/soft nanocomposite films with relatively good magnetic properties. However, for the bulk nanocomposites with complex microstructure, their practical $(BH)_{max}$ is far lower than the theoretically predicted value, remaining a big challenge. Meanwhile, consolidation process is required to prepare full-dense bulk magnets for the practical application of nanostructured materials. Unlike the traditional NdFeB containing low-melting-point Nd-rich phase for the liquid phase sintering and full densification, there is no similar RE-rich phase in the nanocomposite particles. Besides, the conventional high-temperature sintering for long periods may easily lead to abnormal grain growth, destroying the nanostructure required for strong exchange coupling. To solve the above-mentioned difficulties, development of novel approaches to prepare full-dense bulk nanocomposite magnets with effectively inhibited grain growth and maintenance of the nanostructure is necessary.

The bulk nanocomposites are generally categorized into isotropic and anisotropic magnets. For the former, new consolidation techniques mainly include spark plasma sintering (SPS), shock wave compression (SWC), high pressure warm compaction (HPWC), induction heating compaction (IHC) and combustion driven compaction (CDC) (Yue et al., 2017). SPS with the advantages of low sintering temperature, fast heating speed, short sintering time, and pressure sintering inhibits the abnormal grain growth during densification process while maintaining their inherent characteristics. The bulk $Nd_2Fe_{14}B/\alpha$-Fe nanocomposites by SPS possess fine and uniform microstructure for strong exchange-coupling effect, generating the B_r of 0.99 T, H_{ci} of 0.48 T and $(BH)_{max}$ of 12.6 MGOe (Yue et al., 2006). For the SWC method, a variety of explosive loading devices have been employed to rapidly deposit energy into the target powders by shock compression within microsecond duration (Prummer, 1989). Disk-shaped bulk $Pr_2Fe_{14}B/\alpha$-Fe nanocomposite magnets ($\Phi 12 \times 4$ mm^3) have been prepared via the SWC method, exhibiting 99% of the theoretical density, improved B_r of 0.96 T and $(BH)_{max}$ of 13.9 MGOe (Jin et al., 2004). For the HPWC involving high pressure and modest heating, metallic powders exhibit satisfactory plasticity and compressibility to form bulk magnets with high density. The $Sm_2(Fe, Si)_{17}C_x/\alpha$-Fe bulk nanocomposites with average grain size less than 15 nm and $(BH)_{max}$ of 25 MGOe have been fabricated by HPWC technology under 3 GPa pressure (Zhang et al., 2000).

The theoretical $(BH)_{max}$ of anisotropic magnet is four times that of the corresponding isotropic one. Therefore, it is significant to prepare anisotropic nanostructured permanent magnets, despite the the polycrystalline characteristic of the nanocomposites. The bottom-up method, high-pressure compression and severe plastic deformation approaches have been explored for the nanocomposite magnets with oriented hard magnetic phase along the easy axis. Hu et al. have used the bottom-up approach which involves initial powder generation via surfactant-assisted ball

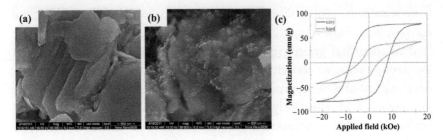

FIGURE 8.3 SEM images of (a) the SmCo$_5$ nanoflakes and (b) hybrid SmCo$_5$/α-Fe powers. (c) Hysteresis loops of the bulk anisotropic SmCo$_5$/α-Fe nanocomposite with 5 wt% α-Fe along the easy and hard axis. Reprinted with permission from (Hu et al., 2012). Copyright (2012) Elsevier.

milling and chemical coating, subsequently magnetic orientation in the metal mold followed by SPS sintering to produce anisotropic bulk SmCo$_5$/α-Fe nanocomposites (Figures 8.3a and b) (Hu et al., 2012). The remarkable magnetization difference along the easy and hard directions demonstrates strong magnetic anisotropy (Figure 8.3c). High remanence ratio also reveals strong exchange coupling effect. The high-pressure compression produces anisotropic bulk nanocomposite magnets by adjusting the growth and orientation of the Nd$_2$Fe$_{14}$B nanocrystals in the amorphous Nd-poor Nd$_9$Fe$_{85}$B$_6$ alloys by high pressure (Wu et al., 2008). Through a multi-field coupling deformation, the oriented rod-like SmCo hard phase and equiaxed Fe(Co) soft phase can be well controlled by stress, strain, and temperature gradient fields (Li et al., 2017).

Nanocomposite permanent magnets are advantageous due to the potentially ultra-high $(BH)_{max}$, high corrosion resistance and low cost, exhibiting broad application prospects. However, most nanocomposite magnets are still in the laboratory research. Particularly, the preparation approaches and other factors including the crystallographic orientation of hard phase, the two-phase grain size, fraction and distribution and interface, have greatly limited the exchange coupling of the magnets, thereby generating a huge disparity between the practical and theoretical $(BH)_{max}$ values. Besides, the coercivity mechanism of nanocomposites also remains unclear till now. With regard to the conventional nucleation-type or pinning-type coercivity mechanism, the situation of nanocomposite becomes more complicated. The introduction of soft phase yields the trade-off between coercivity and remanence, which is also a barrier for large $(BH)_{max}$. To sum up, there are still many challenges and difficulties to overcome for the nanocomposite permanent magnets in the future, from composition design, preparation process, modification of interface structure to fundamental understanding of the coercivity mechanism.

8.2 SmFeN

8.2.1 Crystal Structure

In 1990, nitrogen interstitial compound $RE_2Fe_{17}N_x$ has been fabricated via gas-solid reaction (Coey and Sun, 1990). Unlike the parent compound RE_2Fe_{17}, the $RE_2Fe_{17}N_x$ nitride exhibits strong ferromagnetism. Among various $RE_2Fe_{17}N_x$, the $Sm_2Fe_{17}N_x$ compound exhibits strong c-axis anisotropy and excellent intrinsic magnetic properties at room temperature, making it a potentially attractive permanent magnet. As shown in Figure 8.4, the $Sm_2Fe_{17}N_x$ maintains the same rhombohedral Th_2Zn_{17}-type ($R\bar{3}m$) crystal structure as its Sm_2Fe_{17} precursor (Coey et al., 2019). In each unit cell, there exists three 9e octahedral interstitial sites per formula to accommodate the nitrogen atoms, which explains the saturated nitridation of $x = 3$. The introduction of nitrogen and resultant lattice expansion increases the Fe–Fe distance in the Fe sublattice, strengthening the positive exchange between Fe–Fe atoms and improving the Curie temperature T_C (Yang et al., 1991). Meanwhile, the nitrogen atom in the interstitial position creates strong and negative gradient of electric field on the rare earth element Sm, giving rise to the large and positive magnetocrystalline

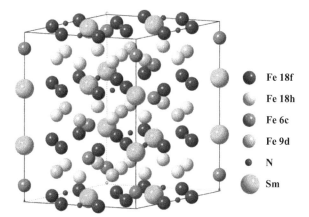

FIGURE 8.4 Crystallographic structure of $Sm_2Fe_{17}N_3$. Reprinted with permission from (Coey et al., 2019). Copyright (2019) Elsevier.

TABLE 8.1 Comparisons on the Theoretical Magnetic Properties of $Nd_2Fe_{14}B$ and $Sm_2Fe_{17}N_x$ Permanent Magnets (Coey et al., 2019)

Materials	K_1 (MJ/m³)	J_s (T)	$(BH)_{max}$ (kJ/m³)	T_C (°C)
$Nd_2Fe_{14}B$	4.9	1.61	515	315
$Sm_2Fe_{17}N_x$	8.6	1.54	475	476

anisotropy constant K_1. As listed in Table 8.1, $Sm_2Fe_{17}N_x$ compound possesses superior K_1 and T_C to $Nd_2Fe_{14}B$, as well as almost equivalent J_s (Coey et al., 2019), serving as a promising permanent material.

8.2.2 Nitriding Mechanism

From the perspective of thermodynamics, the reaction of interstitial nitrogen with Sm_2Fe_{17} metallic compound can be simply described as (Brennan et al., 1995)

$$Sm_2Fe_{17} + \frac{3}{2}N_2 = Sm_2Fe_{17}N_3 \quad (8.1)$$

The resultant $Sm_2Fe_{17}N_3$ is thermodynamically metastable, considering the competitive decomposition reaction following

$$Sm_2Fe_{17}N_3 = 2SmN + Fe_4N + 13\alpha - Fe \quad (8.2)$$

Therefore, the diffusion process should be controlled at a metastable equilibrium state to avoid the decomposition of SmFeN. Meanwhile, the diffusivity of nitrogen in the lattice of Sm_2Fe_{17} host during low-temperature reaction is critical to both magnetic performance and structural stability. Many efforts have been made to establish the kinetic model of the nitriding process of the SmFeN. Under the atmosphere of N_2 or NH_3, nitrogen is adsorbed on the surface of the SmFe alloy powders. The decomposed nitrogen atoms with sufficient energy then enter the lattice gap of the Sm_2Fe_{17} compound at elevated temperatures. The surface region tends to possess a higher nitrogen concentration than that of the core. Upon elongated nitriding time, the concentration gradient of nitrogen yields the diffusion inside the particles until a balanced state is achieved.

Consequently, diffusion of nitrogen in the lattice is the main factor limiting the nitridation. Two different atomic diffusion mechanisms have been proposed, including the free diffusion model and the trapping diffusion model. For the former, the nitrogen atoms diffuse and occupy the 9e octahedral interstitial position in the Th_2Zn_{17} rhombohedral structure (Skomski and Coey, 1993a). Contrastly, for the trapping diffusion model, there exists different interstitial sites with different diffusivities, i.e. the trapped-type atoms are strongly bonded and cannot migrate freely, while the free-type atoms possess weak interaction with the lattice and a larger diffusion constant (Zhang et al., 1997). As revealed by the extreme case of $Y_2Fe_{17}N_x$ system with a small cell volume, N atoms at the 9e sites are trapped and immobilized due to a strong N-lattice interaction, while N atoms at the 18g sites are mobile (Zhang et al., 1997). Diffusion of the nitrogen atoms occurs with the voidal diffusion mechanism, indicating that the migration of interstitial nitrogen requires the unoccupied voids. The 9e voids of the Sm_2Fe_{17} intermetallic compound however, are linked together in the close-packed c plane, as illustrated in Figure 8.5 (Christodoulou and Komada, 1994). The atoms diffusing from void to void should overcome the large energy barrier due to the short interatomic distance and strong lattice bonding with the neighboring Sm/Fe atoms. Consequently, it has been proposed that the diffusion of nitrogen between the 9e sites requires temporary vacant sites as the diffusion pathways, i.e. the 18g tetrahedral sites located out of the c plane. The immobilization induces complete N accommodation at the 9e sites in the outer shell of the SmFeN particle, and the mobile N atoms at the high-energy 18g sites subsequently migrate inward before encountering

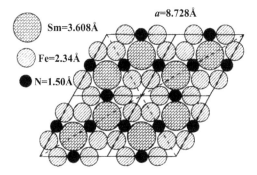

FIGURE 8.5 Close-packed structure of the c plane of the $Sm_2Fe_{17}N_3$ intermetallic compound. Reprinted with permission from (Christodoulou and Komada, 1994). Copyright (1994) Elsevier.

an unoccupied low-energy 9e site to reach the meta-equilibrium state (Christodoulou and Komada, 1995).

8.2.3 Preparation Techniques

The fabrication of SmFeN bulk magnets starts with the preparation of SmFeN powders, which mainly involves rapid quenching (RQ), mechanization (MA), hydrogenation–decomposition–desorption–recombination (HDDR), reduction and diffusion (RD). For the RQ method pioneered by Katter et al. in 1991, SmFe alloy is smelted into an ingot, quenched onto a high-speed copper roller and subjected to gas-solid nitriding treatment at optional annealing conditions (Katter et al., 1991). Daido Steel has utilized the RQ approach to prepare commercial Nitroquench SmFeN powders, which are comparable to the Magnequench isotropic NdFeB bonded magnet. The MA method consists of alloy proportioning, mechanical grinding, heat treatment and nitriding treatment to produce isotropic $Sm_2Fe_{17}N_x$ powders. The HDDR method disproportionates the SmFe alloy powders in hydrogen atmosphere, followed by refining during the dehydrogenation process, hydrogen desorption upon heating, and subsequent recombination step. As a result, the crystal lattice of the SmFe alloy becomes larger and the N atoms can easily diffuse into the lattice. Both the MA and HDDR methods are successful in yielding high-coercivity SmFeN powders up to 3.5 T and 4.4 T, respectively (Muller et al., 1996). For the RD method, raw powders (Sm_2O_3, Fe and Ca) are mechanically mixed, and the Ca reduces the Sm_2O_3 during heat treatment, facilitating the reduced Sm to combine with the Fe to form the Sm_2Fe_{17} (Okada et al., 2017). The $Sm_2Fe_{17}N_3$ powders developed by the calcium thermal RD method give rise to large $(BH)_{max}$ of 46.6 MGOe (Ishikawa et al., 2011) and coercivity of 2.3 T (Hirayama et al., 2016).

The next step is to prepare the SmFeN bulk magnets. The thermodynamically metastable SmFeN may easily decompose into the more stable SmN and α-Fe when heated above ~600 °C, which is destructive to the magnetic performance. As a result, the high-temperature powder metallurgy is excluded to prepare the SmFeN bulk magnets. Alternative consolidation techniques including explosive compaction (Chiba et al., 2007), shock compaction (Chiba et al., 2004), spark plasma sintering (Zhang et al., 2007; Saito, 2008; 2010; 2014) met with limited success for commercial applications. Comparably, the metal bonded magnets with the aid of low-melting-point Zn binder have achieved better results to

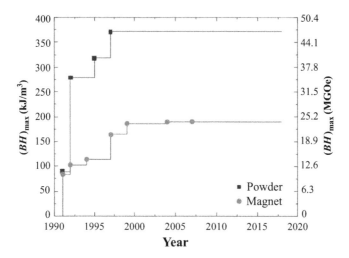

FIGURE 8.6 Progress in $(BH)_{max}$ of SmFeN powders and bulk magnets. Reprinted with permission from (Coey et al., 2019). Copyright (2019) Elsevier.

imitate the high-temperature sintering (Otani et al., 1991; Noguchi et al., 1999; Matsuura et al., 2018), as evidenced by the record-high $(BH)_{max}$ of 23.4 MGOe of the bonded Zn/Sm$_2$(Fe$_{0.9}$Co$_{0.1}$)$_{17}$N$_x$ magnets. To date however, none of the SmFeN-based bulk magnets exhibits high $(BH)_{max}$ above 25 MGOe, which is much lower than the performance of magnetic powders, as shown in Figure 8.6 (Coey et al., 2019).

Commercial SmFeN powders have been mainly produced by companies of Sumitomo, Nichia, and Magvalley (Coey, 2020). Although the production of the SmFeN is less than 1% of the NdFeB, there are prospects of increasing market share due to the reduced material cost and balanced RE utilization. Partial replacement of Fe or Sm by other elements, and minor introduction of other interstitial elements C or H, are possible ways to stabilize the interstitial compounds and enhance the magnetic performance of SmFeN-based bulk magnets.

8.3 Mn-BASED ALLOYS

Among inexpensive 3d metals, Mn with half-filled 3d orbitals exhibits a large magnetic moment (Sato et al., 2020). Mn-based hard magnetic materials potentially surpass the Fe-based or Co-based magnets. However, due to manganese dilemma that the proximity of Mn generates an antiferromagnetic coupling below the Néel temperature of 90 K (Coey, 2014), the α-Mn metal by itself exhibits spontaneous antiferromagnetism. According to the Bethe-Slater rule, the introduction of alloying elements such as Bi, Al, C, Ga, Ge, B, Sb and As can increase the atomic distance of Mn and transforms the antiferromagnetic coupling into ferromagnetic coupling. Among these alloys, the MnBi, MnAl and MnAlC with better performance are considered to be promising gap magnets.

8.3.1 MnBi Alloy

In the MnBi binary systems, the low-temperature phase (LTP) α-MnBi with equiatomic composition $Mn_{50}Bi_{50}$ is ferromagnetic with preferable hard magnetic properties, i.e. M_s of 0.91 T, K_1 of 1.6 MJ/m³ and theoretical $(BH)_{max}$ of 20 MGOe at 300 K, together with T_C of 633 K (Cui et al., 2018; Mohapatra and Liu, 2018). It belongs to the hexagonal $P6_3/mmc$ space group at room temperature ($\alpha = \beta = 90°$, $\gamma = 120°$, $a = b = 4.283$ Å, $c = 6.110$ Å) (Yang et al., 2002). As determined by the neutron diffraction, the moment per Mn reaches 3.5 μ_B at room temperature (Yang et al., 2013). For the high-temperature phase (HTP) β-MnBi with the atomic composition of $Mn_{1.08}Bi$ (Sarkar and Mallick, 2020), partial Mn atoms enter into the interstitial sites and are antiferromagnetically coupled due to the expansion of crystal lattice (Figure 8.7). Consequently, the β-MnBi phase is a non-ferromagnetic material, with the space group of $Pmma$ ($\alpha = \beta = \gamma = 90°$, $a = 5.959$ Å, $b = 4.334$ Å, $c = 7.505$ Å at 630 K) (Cui et al., 2018). A unique feature of α-MnBi is the positive temperature coefficient of coercivity, i.e.

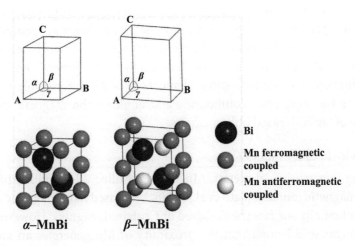

FIGURE 8.7 Crystal structures of the α-MnBi and β-MnBi.

the improved coercivity with increasing temperature (Zarkevich et al., 2014). As reported, the coercivity of MnBi powders reaches 25.8 kOe at 553 K, almost twofold of the value at 300 K (~14 kOe) (Saha et al., 2002). Above characteristics make the MnBi magnet a potential candidate filling the magnetic gap between the RE-free and RE-based permanent magnets for high-temperature applications (Kontos et al., 2020).

From the MnBi phase diagram, the Mn precipitates firstly from the MnBi liquid during cooling (Formula 8.3), then the HTP forms by the peritectic reaction at 719 K (Formula 8.4). With further cooling at 628 K, the HTP reacts with the residual liquid phase and eventually form the LTP (Formula 8.5).

$$L \rightarrow S-Mn+L-Bi \quad (1373K) \tag{8.3}$$

$$S-Mn+L-Bi \rightarrow \beta MnBi \quad (719K) \tag{8.4}$$

$$L-Bi+\beta-MnBi \rightarrow \alpha-MnBi \quad (628\ K) \tag{8.5}$$

Limited by the slow elemental diffusion at low temperature, synthesizing the single-phase LTP on large scale is an extremely difficult task. So far, many efforts have been made to prepare high-purity MnBi magnets, including solid-state reaction sintering (Mitsui et al., 2014), mechanical alloying (Szlaferek and Wrzeciono, 1997), chemical method (Rama Rao et al., 2014; Kirkeminde et al., 2015), as well as melt-spinning (Guo et al., 1990; Yang et al., 2011). Among them, post-casting annealing for prolonged durations are required for the solid-phase reaction and mechanical alloying methods to improve the phase purity. The in-field solid-phase reaction sintering for enhanced purity of MnBi phase (~70 wt%), takes at least 48 h under a large magnetic field of 15 T (Mitsui et al., 2014). The mechanical alloying method even cost 120 h (Szlaferek and Wrzeciono, 1997). Although the chemical method is beneficial to prepare ultrafine crystal grains for improved coercivity, the low phase purity, decreased remanence, together with the high cost of the complex process are the major limitations (Kirkeminde et al., 2015). Comparably, conventional metallurgical routes such as arc melting, melt-spinning, ball milling and heat treatment which are also compatible with industrial production experience continual progress in recent years. The rapid solidification with high cooling rate followed by simple heat treatment yields LTP-MnBi approaching 99% purity (Cui et al., 2014; Jensena et al., 2019), which is suitable as precursor powders for the fabrication of bulk magnets. Low-temperature and low-energy ball milling has been adopted to mitigate the decomposition of LTP, producing over 97% LTP with particle size of 1–5 μm and coercivity of 12 kOe (Nguyen et al., 2014).

Through the in-situ annealing of Bi/Mn/Bi trilayers on the glass substrate, high-purity Mn_xBi_{100-x} ($x = 55, 50, 45$) films with strong c-axis texture has been prepared, generating a record-high $(BH)_{max}$ value of 16.3 MGOe for the optimum $Mn_{50}Bi_{50}$ thin film, as shown in Figure 8.8 (Zhang et al., 2015). For the large-scaled MnBi bulk magnet however, the experimental $(BH)_{max}$ value reported so far is much lower than the theoretical limit. Fully-dense anisotropic MnBi magnet has been prepared by consolidating the oriented powders via hot compaction (593 K for 10 min, 300 MPa, $\leq 4 \times 10^{-5}$ mbar), generating H_{cj} of 0.65 T, B_r of 0.5 T, and $(BH)_{max}$ of 5.8 MGOe at 300 K (Rama Rao et al., 2013). The $(BH)_{max}$ value has been increased to 8.7 MGOe in the bulk form (~1 cm^3 cube) by Cui's group (Poudyal et al., 2016) and 7.3 MGOe (3 cm^3) by Kim's group (Kim et al., 2017). Further improvement of the $(BH)_{max}$ for MnBi bulk magnet requires more in-depth investigations, from the composition design to the processing modifications, or via developing new cost-effective techniques yielding high-purity LTP. Besides, concerns also arise on the high oxidation susceptibility of MnBi and affordable cost of rare Bi. Not only has the MnBi been reported to easily oxidize and decompose to MnO/MnO_2 (Sarkar and Mallick, 2020), but aging experiments also confirm notable M_s decrement of the LTP-MnBi by 54% after 6 days and complete loss of ferromagnetism after 4 months (Villanueva et al., 2019). Since Bi is a byproduct of Pb production with lower crustal abundance (2.5×10^{-6} %) than Ga (1.9×10^{-3} %) (Patel et al., 2018), the increasing demand of Bi for mass production may hinder further applications due to the potentially soaring cost.

8.3.2 MnAl and MnAlC Alloys

The τ-MnAl phase with tetragonal $L1_0$ structure is a hard magnetic material with moderate magnetic properties and low cost. It is also a solid-solution metastable phase with the excessive Mn content ranging between 51 and 58 at% (Cui et al., 2018). The magnetic moment of Mn at the 1a (0, 0, 0) sites are arranged in parallel via ferromagnetic coupling (Park et al., 2010).

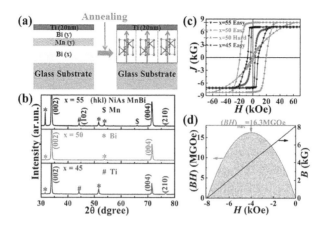

FIGURE 8.8 High-purity Mn_xBi_{100-x} (x = 55, 50, 45) thin films with large $(BH)_{max}$. (a) The schematics of Bi/Mn/Bi trilayer deposition and the formation of texture. (b) X-ray diffraction patterns, and (c) hysteresis loops. (d) (BH)–H curve for the film when x = 50. Reprinted with permission from (Zhang et al., 2015). Copyright (2015) John Wiley & Sons.

However, excessive Mn also enters into the 1d (1/2, 1/2, 1/2) sites to antiferromagnetically couple with the nearest-neighbor Mn atoms at the 1a (0, 0, 0) sites (Zhao et al., 2019), leading to decreased J_s. Theoretically, the MnAl exhibits J_s of 0.75 T, K_1 of 1.7 MJ/m³, T_C of 630 K and $(BH)_{max}$ of 13.2 MGOe (Kontos et al., 2020; Bohlmann et al., 1981). However, the ferromagnetic τ-MnAl is a metastable phase that formed by rapidly cooling of the high-temperature ε phase. It can be easily decomposed into adjacent $γ_2$-Al_8Mn_5 phase and β-Mn phase based on the phase diagram. Minor alloying with the interstitial C has been proven to improve the structural stability of the tetragonal τ-MnAl phase. The C atoms preferentially enter into the (1/2, 1/2, 1/2) position (Pareti et al., 1986), which effectively prevent into the occupation of the Mn atoms for increased J_s. Therefore, the theoretical $(BH)_{max}$ of the MnAlC reaches 16.8 MGOe, but at the sacrifice of K_1 and T_C. Compared to the MnBi, the low-cost MnAl material with ~4 $/kg also possesses higher elastic modulus and improved machinability, which is suitable for applications that do not require high performance but are price-sensitive (Rial et al., 2017; Patel et al., 2018; Mohapatra and Liu, 2018).

Owing to the abundant resources of Mn, Al and C, easy purification and separation, low cost and better magnetic properties than Alnico,

MnAl and MnAlC materials have been extensively studied (Tong et al., 2020; Nguyen et al., 2018). However, since the ordered metastable τ phase is obtained from the nonmagnetic hexagonal ε phase through rapid cooling, the synthesis of MnAl magnet containing a high fraction of τ phase is a strenuous challenge. The transition from ε to τ phase is generally considered with two possibilities, one with orthorhombic ε′ phase as the intermediate phase, the other with the fcc phase (Pareti et al., 1986; Cui et al., 2018). However, both transformations induce a high density of the crystal defects, such as the stacking faults, antiphase boundaries and lattice distortions, which are detrimental to the coercivity of τ-MnAl. Till now, the coercivity values for MnAl and MnAlC are lower than 6.5 kOe at room temperature, except the reported 10.7 kOe in the 11 nm-thick MnAl film epitaxially grown on the GaAs substrates (Nie et al., 2013). For the meltspun ribbons, the coercivity is even lower than 2 kOe. Moreover, with time going by, the magnetism of metastable τ-MnAl phase weakens until it fails. Therefore, more efforts are required to achieve breakthroughs in the stability and performance of MnAl and MnAlC as non-RE candidates for commercial applications.

REFERENCES

Bader, S. D. 2006. Colloquium: opportunities in nanomagnetism. *Reviews of Modern Physics* 78(1):1–15.

Bohlmann, M. A., J. C. Koo, and J. H. Wise. 1981. Mn-Al-C for permanent magnets (invited). *Journal of Applied Physics* 52(3):2542–3.

Brennan, S., R. Skomski, Q. Qi, and J. M. D. Coey. 1995. Is $Sm_2Fe_{17}N_x$ a two-phase system? *Journal of Magnetism and Magnetic Materials* 140–144:999–1000.

Buschow, K. H. J., D. B. De Mooij, and R. Coehoorn. 1988. Metastable ferromagnetic materials for permanent magnets. *Journal of the Less Common Metals* 145:601–11.

Chiba, A., K. Hokamoto, S. Sugimoto, T. Kozuka, A. Mori, and E. Kakimoto. 2007. Explosive consolidation of Sm-Fe-N and Sm-Fe-N/(Ni, Co) magnetic powders. *Journal of Magnetism and Magnetic Materials* 310(2):881–3.

Chiba, A., K. Ooyabu, Y. Morizono, T. Maeda, S. Sugimoto, and T. Kozuka. 2004. Shock consolidation of Sm-Fe-N magnetic powders and magnetic properties. *Materials Science Forum* 449–452:1037–40.

Coehoorn, R., D. B. De Mooij, and C. De Waard. 1989. Meltspun permanent-magnet materials containing Fe_3B as the main phase. *Journal of Magnetism and Magnetic Materials* 80(1):101–4.

Coehoorn, R., D. B. de Mooij, J. P. W. B. Duchateau, and K. H. J. Buschow. 1988. Novel permanent magnetic materials made by rapid quenching. *Journal De Physique* 49(C8):669–70.

Coey, J. M. D. 2014. New Permanent Magnets; Manganese Compounds. *Journal of Physics: Condensed Matter* 26(6):064211.

Coey, J. M. D. 2020. Perspective and prospects for rare earth permanent magnets. *Engineering* 6(2):119–31.

Coey, J. M. D., and H. Sun. 1990. Improved magnetic properties by treatment of iron-based rare earth intermetallic compounds in Ammonia. *Journal of Magnetism and Magnetic Materials* 87(3):L251–L254.

Coey, J. M. D., P. Stamenov, S. B. Porter, M. Venkatesan, R. Zhang, and T. Iriyama. 2019. Sm-Fe-N revisited; Remanence enhancement in melt-spun Nitroquench material. *Journal of Magnetism and Magnetic Materials* 480:186–92.

Christodoulou, C., and N. Komada. 1994. On the atomic diffusion mechanism and diffusivity of nitrogen atoms in Sm_2Fe_{17}. *Journal of Alloys and Compounds* 206(1):1–13.

Christodoulou, C., and N. Komada. 1995. Anisotropic atomic diffusion mechanism of N, C and H into Sm_2Fe_{17}. *Journal of Alloys and Compounds* 222:27–32.

Cui, J., J. P. Choi, E. Polikarpov, M. E. Bowden, W. Xie, G. S. Li, Z. Nie, N. Zarkevich, M. J. Kramer, and D. Johnson. 2014. Effect of composition and heat treatment on MnBi magnetic materials. *Acta Materialia* 79:374–81.

Cui, J., M. Kramer, L. Zhou, F. Liu, A. Gabay, G. Hadjipanayis, B. Balasubramanian, and D. Sellmyer. 2018. Current progress and future challenges in Rare-earth-free permanent magnets. *Acta Materialia* 158:118–37.

Cui, W. B., Y. K. Takahashi, and K. Hono. 2012. $Nd_2Fe_{14}B$/FeCo anisotropic nanocomposite films with a large maximum energy product. *Advanced Materials* 24:6530–5.

Ding, J., P. G. McCormick, and R. Street. 1993. Remanence enhancement in mechanically alloyed isotropic Sm_7Fe_{93}-nitride. *Journal of Magnetism and Magnetic Materials* 124(1–2):1–4.

Fischer, R., T. Leineweber, and H. Kronmüller. 1998. Fundamental magnetization processes in nanoscaled composite permanent magnets. *Physical Review B* 57(17):10723–32.

Fukunaga, H., R. Horikawa, M. Nakano, T. Yanai, T. Fukuzaki, and K. Abe. 2013. Computer simulations of the magnetic properties of Sm-Co/alpha-Fe nanocomposite magnets with a core-shell structure. *IEEE Transactions on Magnetics* 49(7):3240–3.

Guo, X., A. Zaluska, Z. Altounian, and J. O. Strom-Olsen. 1990. Formation of single-phase equiatomic MnBi by rapid solidification. *Journal of Materials Research* 5(11):2646–51.

Hirayama, Y., A. K. Panda, T. Ohkubo, and K. Hono. 2016. High coercivity $Sm_2Fe_{17}N_3$ submicron size powder prepared by polymerized-complex and reduction-diffusion process. *Scripta Materialia* 120:27–30.

Hu, D. W., M. Yue, J. H. Zuo, R. Pan, D. T. Zhang, W. Q. Liu, J. X. Zhang, Z. H. Guo, and W. Li. 2012. Structure and magnetic properties of bulk anisotropic

SmCo$_5$/alpha-Fe nanocomposite permanent magnets prepared via a bottom-up approach. *Journal of Alloys and Compounds* 538:173–6.

Ishikawa, T., K. Yokosawa, K. Watanabe, and K. Ohmori. 2011. Modified process for high-performance anisotropic Sm$_2$Fe$_{17}$N$_3$ magnet powder. *Journal of Physics: Conference Series* 266:012033.

Jensena, B. A., W. Tang, X. Liu, A. I. Noltea, G. Ouyang, K. W. Dennis, and J. Cui. 2019. Optimizing composition in MnBi permanent magnet alloys. *Acta Materialia* 181:595–602.

Jin, Z. Q., K. H. Chen, J. Li, H. Zeng, S. F. Cheng, J. P. Liu, Z. L. Wang, and N. N. Thadhani. 2004. Shock compression response of magnetic nanocomposite powders. *Acta Materialia* 52(8):2147–54.

Katter, M., J. Wecker, and L. Schultz. 1991. Structural and hard magnetic properties of rapidly solidified Sm-Fe-N. *Journal of Applied Physics* 70(6):3188–96.

Kim, S., H. Moon, H. Jung, S. Kim, H. Lee, H. Choi-Yim, and W. Lee. 2017. Magnetic properties of large-scaled MnBi bulk magnets. *Journal of Alloys and Compounds* 708:1245–9.

Kirkeminde, A., J. Shen, M. G. Gong, J. Cui, and S. Q. Ren. 2015. Metal-redox synthesis of MnBi hard magnetic nanoparticles. *Chemistry of Materials* 27(13):4677–81.

Kneller, E. F., and R. Hawig. 1991. The exchange-spring magnet: a new material principle for permanent magnets. *IEEE Transactions on Magnetics* 27(4):3588–600.

Kontos. S., A. Ibrayeva, J. Leijon, G. Mörée, A. E. Frost, L. Schönström, K. Gunnarsson, P. Svedlindh, M. Leijon, and S. Eriksson. 2020. An overview of MnAl permanent magnets with a study on their potential in electrical machines. *Energies* 13(21):5549.

Li, H. L., X. H. Li, D. F. Guo, L. Lou, W. Li, and X. Y. Zhang. 2016. Three-dimensional self-assembly of core/shell-like nanostructures for high-performance nanocomposite permanent magnets. *Nano Letters* 16(9):5631–8.

Li, X. H., L. Lou, W. P. Song, G. W. Huang, F. C. Hou, Q. Zhang, H. T. Zhang, J. W. Xiao, B. Wen, and X. Y. Zhang. 2017. Novel bimorphological anisotropic bulk nanocomposite materials with high energy products. *Advanced Materials* 29(16):1606430.

Liu, F., Y. L. Hou, and S. Gao. 2014. Exchange-coupled nanocomposites: chemical synthesis, characterization and applications. *Chemical Society Reviews* 43(23):8098–113.

Liu, J. P. 2009. *Exchange-coupled nanocomposite permanent magnets, nanoscale magnetic materials and applications*. New York: Springer Science and Business Media. LLC.

Liu, W., Z. D. Zhang, J. P. Liu, L. J. Chen, L. L. He, Y. Liu, X. K. Sun, and D. J. Sellmyer. 2002. Exchange coupling and remanence enhancement in nanocomposite multilayer magnets. *Advanced Materials* 14(24):1832–4.

Manaf, A., R. A. Buckley, and H. A. Davies. 1993. New nanocrystalline high-remanence Nd-Fe-B alloys by rapid solidification. *Journal of Magnetism and Magnetic Materials* 128(3):302–6.

Matsuura. M., T. Shiraiwa, N. Tezuka, S. Sugimoto, T. Shoji, N. Sakuma, and K. Haga. 2018. High coercive Zn-bonded Sm-Fe-N magnets prepared using fine Zn particles with low oxygen content. *Journal of Magnetism and Magnetic Materials* 452:243–8.

Mitsui, Y., R. Y. Umetsu, K. Koyama, and K. Watanabe. 2014. Magnetic-field-induced enhancement for synthesizing ferromagnetic MnBi phase by solid-state reaction sintering. *Journal of Alloys and Compounds* 615:131–4.

Mohapatra, J., and J. P. Liu. 2018. Rare-earth-free permanent magnets: the past and future. *Handbook of Magnetic Materials* 27:1–57.

Muller, K. H., L. Cao, N. M. Dempsey, and P. A. P. Wendhausen. 1996. Sm_2Fe_{17} interstitial magnets (invited). *Journal of Applied Physics* 79(8):5045–50.

Neu, V., S. Sawatzki, M. Kopte, Ch Mickel, and L. Schultz. 2012. Fully epitaxial, exchange coupled $SmCo_5$/Fe multilayers with energy densities above 400 kJ/m^3. *IEEE Transactions on Magnetics* 48(11):3599–602.

Nguyen, V. T., F. Calvayrac, A. Bajorek, and N. Randrianantoandro. 2018. Mechanical alloying and theoretical studies of MnAl(C) magnets. *Journal of Magnetism and Magnetic Materials* 462:96–104.

Nguyen, V. V., N. Poudyal, X. B. Liu, J. Ping Liu, K. Sun, M. J. Kramer, and J. Cui. 2014. Novel processing of high-performance MnBi magnets. *Materials Research Express* 1(3):036108.

Nie, S. H., L. J. Zhu, J. Lu, D. Pan, H. L. Wang, X. Z. Yu, J. X. Xiao, and J. H. Zhao. 2013. Perpendicularly magnetized τ-MnAl (001) thin films epitaxied on GaAs. *Applied Physics Letters* 102(15):152405.

Noguchi, K., K. Machida, K. Yamamoto, M. Nishimura, and G. Adachi. 1999. High-performance resin-bonded magnets produced from zinc metal-coated $Sm_2(Fe_{0.9}Co_{0.1})_{17}N_x$ fine powders. *Applied Physics Letters* 75(11):1601–3.

Okada, S., K. Suzuki, E. Node, K. Takagi, K. Ozaki, and Y. Enokido. 2017. Preparation of submicron-sized $Sm_2Fe_{17}N_3$ fine powder with high coercivity by reduction-diffusion process. *Journal of Alloys and Compounds* 695:1617–23.

Otani, Y., A. Moukarika, H. Sun, and J. M. D. Coey. 1991. Metal bonded $Sm_2Fe_{17}N_{3-\delta}$ magnets. *Journal of Applied Physics* 69(9):6735–7.

Pareti, L., F. Bolzoni, F. Leccabue, and A. E. Ermakov. 1986. Magnetic anisotropy of MnAl and MnAlC permanent magnet materials. *Journal of Applied Physics* 59(11):3824–8.

Park, J. H., Y. K. Hong, S. Bae, J. J. Lee, J. Jalli, G. S. Abo, N. Neveu, S. G. Kim, C. J. Choi, and J. G. Lee. 2010. Saturation magnetization and crystalline anisotropy calculations for MnAl permanent magnet. *Journal of Applied Physics* 107(9):09A731.

Patel, K., J. M. Zhang, and S. Q. Ren. 2018. Rare-earth-free high energy Product manganese-based magnetic materials. *Nanoscale* 10(25):11701–18.

Poudyal, N., X. B. Liu, W. Wang, V. V. Nguyen, Y. L. Ma, K. Gandha, K. Elkins, J. P. Liu, K. W. Sun, M. J. Kramer, and J. Cui. 2016. Processing of MnBi bulk magnets with enhanced energy product. *AIP Advances* 6(5):056004.

Prummer, R. 1989. Explosive compaction of powders, principle and prospects. *Materialwissenschaft Und Werkstofftechnik* 20(12):410–15.

Rama Rao, N. V., A. M. Gabay, and G. C. Hadjipanayis. 2013. Anisotropic fully dense MnBi permanent magnet with high energy product and high coercivity at elevated temperatures. *Journal of Physics D: Applied Physics* 46(6):062001.

Rama Rao, N. V., A. M. Gabay, X. Hu, and G. C. Hadjipanayis. 2014. Fabrication of anisotropic MnBi nanoparticles by mechanochemical process. *Journal of Alloys and Compounds* 586:349–52.

Rial, J., M. Villanueva, E. Céspedes, N. López, J. Camarero, L. G. Marshall, L. H. Lewis, and A. Bollero. 2017. Application of a novel flash-milling procedure for coercivity development in nanocrystalline MnAl permanent magnet powders. *Journal of Physics D: Applied Physics* 50(10):105004.

Rong, C. B., H. W. Zhang, R. J. Chen, S. L. He, and B. G. Shen. 2006. The role of dipolar interaction in nanocomposite permanent magnets. *Journal of Magnetism and Magnetic Materials* 302(1):126–36.

Sabiryanov, R. F., and S. S. Jaswal. 1998a. Electronic structure and magnetic properties of hard/soft multilayers. *Journal of Magnetism and Magnetic Materials* 177:989–90.

Sabiryanov, R. F., and S. S. Jaswal. 1998b. Magnetic properties of hard/soft composites: $SmCo_5/Co_{1-x}Fe_x$. *Physical Review B* 58(18):12071–4.

Saha, S., R. T. Obermyer, B. J. Zande, V. K. Chandhok, S. Simizu, and S. G. Sankar, and J. A. Horton. 2002. Magnetic properties of the low-temperature phase of MnBi, *Journal of Applied Physics* 91(10):8525–7.

Saito, T. 2008. Consolidation of Sm_5Fe_{17} powder by spark plasma sintering method. *Materials Science and Engineering: B* 150(1):38–42.

Saito, T. 2010. Magnetic properties of Sm-Fe-N anisotropic magnets produced by magnetic-field-assisted spark plasma sintering. *Materials Science and Engineering: B* 167(2):75–9.

Saito, T. 2014. Production of Sm–Fe–N bulk magnets by spark plasma sintering method. *Journal of Magnetism and Magnetic Materials* 369:184–8.

Sarkar, A., and A. B. Mallick. 2020. Synthesizing the hard magnetic low-temperature phase of MnBi alloy: challenges and prospects. *JOM* 72(8):2812–25.

Sato, S., S. Irie, Y. Nagamine, T. Miyazaki, and Y. Umeda. 2020. Antiferromagnetism in perfectly ordered $L1_0$-MnAl with stoichiometric composition and its mechanism. *Scientific Reports* 10(1):12489.

Schrefl, T., R. Fischer, J. Fidler, and H. Kronmuller. 1994. Two- and three-dimensional calculation of remanence enhancement of rare-earth based composite magnets (invited). *Journal of Applied Physics* 76(10):7053–8

Skomski, R., and J. M. D. Coey. 1993a. Nitrogen diffusion in Sm_2Fe_{17} and local elastic and magnetic properties. *Journal of Applied Physics* 73(11):7602–11.

Skomski, R., and J. M. D. Coey. 1993b. Giant energy product in nanostructured two-phase magnets. *Physical Review B* 48(21):15812–6.

Skomski, R., and J. M. D. Coey. 1993c. Nucleation field and energy product of aligned 2-phase magnets—progress towards the '1 MJ/M^3' magnet. *IEEE Transactions on Magnetics* 29(6):2860–2.

Sun, S. H., C. B. Murray, D. Weller, L. Folks, and A. Moser. 2000. Monodisperse FePt nanoparticles and ferromagnetic FePt nanocrystal superlattices. *Science* 287(5460):1989–92.

Szlaferek, A., and A. Wrzeciono. 1997. Formation of MnBi by mechanical alloying. *Acta Physica Polonica A* 92(2):315–8.

Tong, X., P. Sharma, and A. Makino. 2020. Investigations on low energy product of MnAl magnets through recoil curves. *Journal of Physics D: Applied Physics* 53(17):175001.

Villanueva, M., C. Navío, E. Céspedes, F. Mompeán, M. García-Hernández, J. Camarero, and A. Bollero. 2019. MnBi thin films for high temperature permanent magnet applications, *AIP Advances* 9(3):035325.

Wu, W., W. Li, H. Y. Sun, H. Li, X. H. Li, B. T. Liu, and X. Y. Zhang. 2008. Pressure-induced preferential growth of nanocrystals in amorphous $Nd_9Fe_{85}B_6$. *Nanotechnology* 19(28):285603.

Yang, J. B., W. B. Yelon, W. J. James, Q. Cai, S. Roy, and N. Ali. 2002. Structure and magnetic properties of the MnBi low temperature phase, *Journal of Applied Physics* 91(10):7866–8.

Yang, J. B., Y. B. Yang, X. G. Chen, X. B. Ma, J. Z. Han, Y. C. Yang, S. Guo, A. R. Yan, Q. Z. Huang, M. M. Wu, and D. F. Chen. 2011. Anisotropic nanocrystalline MnBi with high coercivity at high temperature. *Applied Physics Letters* 99(8):082505.

Yang, Y. B., X. G. Chen, S. Guo, A. R. Yan, Q. Z. Huang, M. M. Wu, D. F. Chen, Y. C. Yang, and J. B. Yang. 2013. Temperature dependences of structure and coercivity for melt-spun MnBi compound. *Journal of Magnetism and Magnetic Materials* 330:106–10.

Yang, Y. C., X. D. Zhang, S. L. Ge, L. S. Kong, and Q. Pan. 1991. Structural and magnetic properties of the new type of rare earth-iron-Nitrogen intermetallic compounds. *Journal of Rare Earths* 9:81–4.

Yue, M., J. X. Zhang, M. Tian, and X. B. Liu. 2006. Microstructure and magnetic properties of isotropic bulk $Nd_xFe_{94-x}B_6$ (x = 6,8,10) nanocomposite magnets prepared by spark plasma sintering. *Journal of Applied Physics* 99(8):08B502.

Yue, M., X. Y. Zhang, and J. P. Liu. 2017. Fabrication of bulk nanostructured permanent magnets with high energy density: challenges and approaches. *Nanoscale* 9(11):3674–97.

Zarkevich, N. A., L. L. Wang, and D. D. Johnson. 2014. Anomalous magnetostructural behavior of MnBi explained: a path towards an improved permanent magnet. *APL Materials* 2(3):032103.

Zeng, H., J. Li, J. P. Liu, Z. L. Wang, and S. H. Sun. 2002. Exchange-coupled nanocomposite magnets by nanoparticle self-assembly. *Nature* 420(6914):395–8.

Zhang, D. T., M. Yue, and J. X. Zhang. 2007. Study on bulk $Sm_2Fe_{17}N_x$ sintered magnets prepared by spark plasma sintering. *Powder Metallurgy* 50(3):215–8.

Zhang, W. Y., P. Kharel, S. Valloppilly, L. P. Yue, D. J. Sellmyer. 2015. High-energy-product MnBi films with controllable anisotropy. *Physica Status Solidi (B)* 252(9):1934–9.

Zhang, X. Y., J. W. Zhang, and W. K. Wang. 2000. A novel route for the preparation of nanocomposite magnets. *Advanced Materials* 12(19):1441–4.

Zhang, Y. D., J. I. Budnick, W. A. Hines, N. X. Shen, and J. M. Gromek. 1997. Nitrogen diffusion in R_2Fe_{17} lattice: a trapping diffusion process. *Journal of Physics: Condensed Matter* 9(6):1201–16.

Zhao, S., Y. Y. Wu, Z. Y. Jiao, Y. X. Jia, Y. C. Xu, J. M. Wang, T. L. Zhang, and C. B. Jiang. 2019. Evolution of intrinsic magnetic properties in L_{10} Mn-Al Alloys doped with substitutional atoms and correlated mechanism: experimental and theoretical studies. *Physical Review Applied* 11(6):064008.

FURTHER READING

Zeng, H., J. Li, J. P. Liu, Z. L. Wang, and S. H. Sun. 2002. Exchange-coupled nanocomposite magnets by nanoparticle self-assembly. *Nature* 420(6914):395–8.

III

Soft Magnetic Materials

CHAPTER 9

Introduction to Soft Magnetic Materials

9.1 APPLICATIONS OF SOFT MAGNETIC MATERIALS

Ferromagnetic materials can be classified as hard and soft magnets based on their coercivity. For the hard magnetic materials, a large external field is usually necessary to saturate the magnetization or to quench its induction. On the contrast, the soft magnetic materials can be both easily saturated and demagnetized. Soft magnetic materials are featured with large magnetization, high permeability and low coercivity, determining that they are sensitive to any applied magnetic field with rapid response. Consequently, soft magnetic materials find important applications for electrical energy conversion as in electric transformers, inductors and motors, etc. (Silveyra et al., 2018).

For power electronics, soft magnets tend to be used as magnetic cores which are the key units of transformers and inductors. The magnetic cores guide and concentrate the flux, and their performance directly determines the efficiency of the transformers (Cha et al., 2009; Pei et al., 2019). As for inductors, when a magnetic core is inserted into an air core inductor, the inductance can be thousands of times larger with the same geometry dimensions and turn number (Fang et al., 2010). For electrical machines, soft magnetic materials are essential to convert electrical energy to mechanical energy in various motors that are widely used in manufactory, transportation and household appliances (Gutfleisch et al., 2011).

DOI: 10.1201/9781003216346-12

9.2 PERFORMANCE REQUIREMENTS FOR SOFT MAGNETIC MATERIALS

The soft magnets are usually required to possess large saturation magnetic induction B_s, low coercivity H_c, high permeability μ and low loss P_c. The B_s determines the power conversion efficiency of the material which is critical for device miniaturization. It is an intrinsic parameter mainly determined by the composition of the material. Since soft magnetic materials need to respond to changes of magnetic field rapidly, low H_c is necessary. The H_c is not only determined by the composition-tuned magnetocrystalline anisotropy constant K_1 and magnetostrictive coefficient λ_s, but also depends on extrinsic factors such as microstructure and stress. On one hand, K_1 and λ_s may be simultaneously reduced to zero for minimized H_c with proper design of the material composition. On the other hand, reduction of grain boundaries and internal stress which hinder the movement of the domain walls are beneficial to decrease the H_c. The μ is positively corelated to the B and inversely corelated to the H. Consequently, methods to enhance the B_s and to reduce the H_c are also applicable to improve the μ. Since soft magnetic materials are mostly used under alternating fields, magnetic loss caused by dynamic magnetization cannot be ignored, which mainly involves the hysteresis loss P_h, eddy current loss P_e and residual loss P_r as introduced in Chapter 3. The P_h and P_r are related to magnetic properties such as magnetocrystalline anisotropy, magnetostriction and the domain structure, which should be considered comprehensively, whereas the P_e can usually be decreased by improving the electrical resistivity ρ.

9.3 DEVELOPMENT OF SOFT MAGNETIC MATERIALS

To fulfill the requirements discussed above, various soft magnetic materials have been developed, which can be categorized into soft magnetic alloys, ferrites and soft magnetic composites. Comparisons of the B_s and H_c for various soft magnetic materials are illustrated in Figure 9.1. Before the 1940s, soft magnetic alloys including silicon steels (FeSi) (Barrett et al., 1900), permalloys (FeNi) (Arnold and Elmen, 1923) and the 'Sendust' alloy (FeSiAl) (Masumoto and Yamamoto, 1937) dominate the applications. Such alloys exhibit high B_s, μ and T_C but low ρ (~10^{-7}–10^{-5} $\Omega\cdot$m) (Fiorillo et al., 1999). This leads to dramatically increased eddy current loss, limiting their high-frequency applications.

Introduction to Soft Magnetic Materials ▪ 155

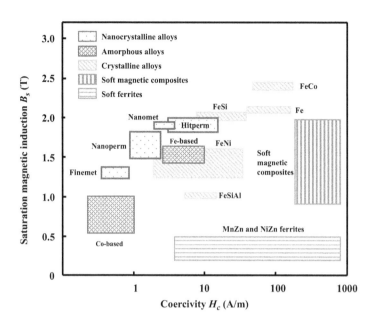

FIGURE 9.1 Magnetic performance of different soft magnetic materials.

Later on, Fe- and Co-based amorphous alloys have been developed, exhibiting increased electrical resistivity of $1.2 \times 10^{-6} – 1.4 \times 10^{-6}$ Ω·m (Fiorillo et al., 1999) due to the addition of various non-metallic atoms to enhance the glass-forming ability as well as the scattering of electrons in the amorphous structure. Compared with the crystalline alloys, reduced H_c can be achieved for the amorphous magnetic alloys as shown in Figure 9.1, resulted from the lack of magnetocrystalline anisotropy. In 1988, a new magnetic alloy FeCuNbSiB with nanocrystalline feature was developed via appropriate heat treatment of the Fe-based amorphous alloy (Yoshizawa et al., 1988). The dispersive growth of nanograins along random crystallographic directions in the amorphous matrix not only gives rise to canceled-out anisotropy for reduced H_c, but also enhanced magnetic coupling for enlarged B_s of ~1.21 T. Moreover, relatively large ρ around 1.2×10^{-6} Ω·m can be maintained (Gutfleisch et al., 2011). The FeCuNbSiB has soon been commercialized as Finemet and developed as one of the most widely used nanocrystalline alloys. After the invention of Finemet, a variety of nanocrystalline systems emerge including the Hitperm (FeCoMBCu, M = Zr, Nb, Hf) (Willard et al., 1999), Nanoperm (FeMBCu, M = Zr, Nb, Hf) (Suzuki et al., 2007) and Nanomet (FeSiBPCu) (Makino et al., 2009).

Performances of these nanocrystalline systems are also shown in Figure 9.1 where the Nanoperm, Hitperm and Nanomet exhibit higher B_s and H_c but lower μ than those of the Finemet. This can be attributed to their compositional difference, mainly the content of the magnetic element Fe and Co, together with the possible formation of Fe–Co pairs for enlarged B_s and H_c as in the Hitperm.

For high-frequency applications, the development of soft magnetic ferrites with resistivity between $10–10^6$ $\Omega \cdot$m soon becomes an alternative solution instead of the alloys (Wu et al., 1980). The soft spinal ferrites are mainly involved with applications at the MHz range (Stoppels, 1996). Nowadays, the most commonly used ferrites are the MnZn and NiZn series. The MnZn ferrites exhibit large initial permeability and are the most important soft magnetic materials for transformer cores (Otsuki et al., 1991), while the NiZn ferrites tend to be used for high-frequency filter inductors (Su et al., 2008; Fukuda et al., 2003; Tatsuya, 2000). The emergence of ferrites has successfully supported the progressive increased frequency for electronic devices due to the low loss. Their B_s, however, may only be around 1/4 of that of the Fe-based soft magnetic alloys (Figure 9.1), since they in fact belong to ferrimagnetic materials. Hence, it is necessary to develop new soft magnets for large-power applications.

Soft magnetic composites (SMCs) have become a fast-developing branch and are replacing some ferrites in the frequency range of 20 kHz to 100 kHz, or even to 1 MHz. The SMCs are made of metallic soft magnetic powders surrounded by insulation layers. Compared with the ferrites, larger magnetic flux density and energy storage capacity can be achieved for the SMCs. Meanwhile, they overcome the major disadvantage of high power loss for the magnetic alloys due to the separation of the magnetic powders by the insulation coatings. Furthermore, the SMCs exhibit isotropic magnetic behavior and can be processed into complex structure via powder metallurgy (Shokrollahi and Janghorban, 2007). Consequently, the SMCs with maintained high B_s (Figure 9.1) and large ρ provide a feasible solution for simultaneous high-power and high-frequency applications, filling the gap between soft magnetic alloys and ferrites.

The development of SMCs can be dated back to the 19th century and exhibits close relationship with the invention of new soft magnetic alloys. At the end of the 19th century, the Fe SMCs were invented simply by wrapping the Fe powders in wax as an insulating agent. This was followed by using carbonyl iron powders as the main component (Birks, 1948). After the invention of other soft magnetic alloy systems including the FeSi, FeNi and FeSiAl, their corresponding SMCs have been produced. Later on, Fe-based amorphous (FeSiB) (Kim and Kim, 2006) and nanocrystalline (FeCuNbSiB) (Jiang et al., 2020) alloys have also exhibited great potential to be fabricated into SMCs.

In summary, soft magnets are technologically important for energy conversion. Along with the requirements for large power, fast response and low loss of related devices, soft magnets including alloys, ferrites and SMCs have been developed. The metallic soft magnets exhibit the largest magnetization while the ferrites are suitable for high-frequency applications due to significantly reduced eddy current loss. The development of SMCs combines the advantages of the two, which fill the application gap with maintained magnetization and reduced loss. Details of each category of the soft magnetic materials will be provided in the following chapters.

REFERENCES

Arnold, H. D., and G. W. Elmen. 1923. Permalloy, an alloy of remarkable magnetic properties. *Journal of the Franklin Institute* 195:621–32.

Barrett, W. F., W. Brown, and R. A. Hadfield. 1900. Electrical conductivity and magnetic permeability of various alloys of Fe. *Science Transaction Royal Dublin Society* 7:67–126.

Birks, J. B. 1948. Microwave magnetic dispersion in carbonyl iron powder. *Physical Review* 74:843–4.

Cha, H. R., S. H. Lee, S. K. Jeon, J. Cho, S. Kikuchi, and K. Nakamura. 2009. Motor core fabrication through ultrasonic vibration compaction of soft magnetic composite. *Japanese Journal of Applied Physics* 48:07GM18.

Fang, D. M., Q. Yuan, X. H. Li, and H. X. Zhang. 2010. Electrostatically driven tunable radio frequency inductor. *Microsystem Technologies* 16:2119–22.

Fiorillo, F., G. Bertotti, C. Appino, and M. Pasquale. 1999. *Soft magnetic materials*. Philadelphia: John Wiley & Sons.

Fukuda, Y., T. Inoue, T. Mizoguchi, S. Yatabe, and Y. Tachi. 2003. Planar inductor with ferrite layers for DC-DC converter. *IEEE Transactions on Magnetics* 39:2057–61.

Gutfleisch, O., M. A. Willard, E. Bruck, C. H. Chen, S. G. Sankar, and J. P. Liu. 2011. Magnetic materials and devices for the 21st century: stronger, lighter, and more energy efficient. *Advanced Materials* 23:821–42.

Jiang, C. Q., X. R. Li, S. S. Ghosh, H. Zhao, Y. F. Shen, and T. Long. 2020. Nanocrystalline powder cores for high-power high-frequency applications. *IEEE Transactions on Power Electronics* 35:10821–30.

Kim, Y. B., and K. Y. Kim. 2006. Effects of the addition of permalloy powder on the high-frequency magnetic properties of Fe-based amorphous powder cores. *IEEE Transactions on Magnetics* 42:2802–4.

Makino, A., H. Men, T. Kubota, K. Yubuta, and A. Inoue. 2009. FeSiBPCu nanocrystalline soft magnetic alloys with high B_s of 1.9 tesla produced by crystallizing hetero-amorphous phase. *Materials Transactions* 50:204–9.

Masumoto, H., and T. Yamamoto. 1937. On a new alloy "Sendust" and its magnetic and electric properties Fe-Si-Al. *Transactions of the Japan Institute of Metals* 1:127–35.

Otsuki, E., S. Yamada, T. Otsuka, K. Shoji, and T. Sato. 1991. Microstructure and physical properties of Mn-Zn ferrites for high-frequency power supplies. *Journal of Applied Physics* 69:5942.

Pei, X. Z., A. C. Smith, L. Vandenbossche, and J. Rens. 2019. Magnetic characterization of soft magnetic cores at cryogenic temperatures. *IEEE Transactions on Applied Superconductivity* 29:1–6.

Shokrollahi, H., and K. Janghorban. 2007. Soft magnetic composite materials (SMCs). *Materials Processing Technology* 189:1–12.

Silveyra, J. M., E. Ferrara, D. L. Huber, and T. C. Monson. 2018. Soft magnetic materials for a sustainable and electrified world. *Science* 362:eaao0195.

Stoppels, D. 1996. Developments in soft magnetic power ferrites. *Journal of Magnetism and Magnetic Materials* 160:323–8.

Su, H., H. W. Zhang, X. L. Tang, and Y. I. Jing. 2008. Influence of microstructure on permeability dispersion and power loss of NiZn ferrite. *Journal of Applied Physics* 103:093903.

Suzuki, K., N. Kataoka, A. Inoue, A. Makino, and T. Masumoto. 2007. High saturation magnetization and soft magnetic properties of bcc Fe-Zr-B alloys with ultrafine grain structure. *Materials Transactions, JIM* 31:743–6.

Tatsuya, N. 2000. Snoek's limit in high-frequency permeability of polycrystalline Ni-Zn, Mg-Zn, and Ni-Zn-Cu spinel ferrites. *Journal of Applied Physics* 88:348.

Willard, M. A., M. Q. Huang, D. E. Laughlin, and M. E. McHenry. 1999. Magnetic properties of HITPERM $(FeCo)_{88}Zr_7B_4Cu_1$ magnet. *Journal of Applied Physics* 85:4421.

Wu, L., T. S. Wu, and C. C. Wei. 1980. Effect of various substitutions on the DC resistivity of ferrites. *Journal of Physics D: Applied Physics* 13:259–66.

Yoshizawa, Y., S. Oguma, and K. Yamauchi. 1988. New Fe-based soft magnetic alloys composed of ultrafine grain structure. *Journal of Applied Physics* 64:6044.

FURTHER READING

Silveyra, J. M., E. Ferrara, D. L. Huber, and T. C. Monson. 2018. Soft magnetic materials for a sustainable and electrified world. *Science* 362:eaao0195.

CHAPTER **10**

Soft Magnetic Alloys

10.1 CRYSTALLINE MAGNETIC ALLOYS

The crystalline magnetic alloys can be roughly divided into FeSi-based alloys with the largest production and FeNi-based systems with zero magnetocrystalline anisotropy constant K_1 or saturation magnetostriction λ_s, which will be discussed in the following sections.

10.1.1 FeSi-Based Magnetic Alloys

The earliest used metallic crystalline soft magnetic material is simply the Fe. Despite the large saturation magnetic induction B_s of the Fe (~2.16 T), its small electrical resistivity ρ (~1.1 × 10^{-7} Ω·m) tends to result in large eddy current loss (Gutfleisch et al., 2011). To overcome this issue, 6.5 wt% Si has been added to form solid solution in the α-Fe phase for increased ρ (~8.2 × 10^{-7} Ω·m) (Ouyang et al., 2019). More importantly, the introduction of Si reduces the magnetocrystalline anisotropy constant K_1 (~20 kJ/m³) and saturation magnetostriction λ_s (~10^{-6}) for reduced coercivity H_c (Hall, 1959; Carr and Smoluchowski, 1951). Relatively large B_s (~1.82 T) can be maintained due to the low content of the non-magnetic Si (Ouyang et al., 2019). Based on the FeSi binary alloy system, a third element, Al has been added and the content of Si (9.6 wt%) and Al (5.4 wt%) have been adjusted to achieve simultaneously zero K_1 and λ_s. The FeSiAl accounts for an important category of magnetic alloys and is also known as 'Sendust' named after its invention place. The increased amounts of non-magnetic Si and Al sacrifice the B_s (~1.05 T), but brings significantly reduced H_c (~1.6 A/m) together with increased initial permeability μ_i (~3.5 × 10^4) and ρ (~8.2 ×10^{-6} Ω·m) (Wakiyama et al., 1981; Masumoto and Yamamoto, 1937). Similarly, the introduction of other elements such as Mn (Narita and Enokizon, 1978) and Cr (Sato-Turtelli et al., 2005) into the FeSi are beneficial to reduce the H_c and enhance the ρ.

Effects of ferromagnetic elements including Co and Ni are two-fold. On one hand, similar to introduction of other transitional metal elements, Co and Ni are beneficial to adjust the K_1 and λ_s. With delicately designed composition for the ternary or quaternary systems, it is possible to achieve simultaneous zero K_1 and λ_s (Hayakawa et al., 1987). On the other hand, large B_s can be maintained or even improved since it depends on the number of electrons per atom, the atomic distance and the orbital radius of the 3d or 4f electrons (Bozorth, 1951). The famous Slater-Pauling curve serves as an excellent reference for changes in the magnetization of binary alloy systems, among which the FeCo alloy exhibits the highest B_s (~2.45 T) with a Co content of approximately 35.0 at% (Sourmail, 2005). Similarly, appropriate addition of Ni is beneficial to slightly improve the B_s. Although the optimal ratio between Fe and Co or Ni may vary for the multi-element alloy systems compared with the binary ones, the Slater-Pauling curve provides useful information on the evolution of B_s via alloying.

10.1.2 FeNi-Based Magnetic Alloys

Compared with the FeSi-based magnetic alloys, the FeNi-based alloys features high permeability and cost due to the existence of large content of Ni (typically from 30.0 wt% to 90.0 wt% for the FeNi permalloys) (Arnold and Elmen, 1923). The K_1 and λ_s reach zero when the Ni content is 76.0 wt% and 80.0 wt%, respectively (Fiorillo et al., 1999). The largest μ of the FeNi can be achieved with a Ni content of 78.5 wt% as both K_1 and λ_s approach zero. Its B_s (~0.75 T), however, is much lower compared with the alloy containing 50.0 wt% Ni (~1.62 T).

Based on the FeNi permalloys, FeNiM (M = Mo, V, W or Nb) alloys have been developed (Masumoto and Hinai, 1974; Rassmann and Hofmann, 1968). Among these alloys, FeNiMo, known as the supermalloy has been used in military applications due to the μ_{max} of 1.0 × 10^5–8.0 × 10^5 and low H_c of 0.3–0.6 A/m at the optimal composition with 78.0 at% Ni and 5.0 at% Mo (Gutfleisch et al., 2011). Other elements such as V, W and Nb have also been added to form the FeNiM alloys, whose role is to tailor K_1 and λ_s to approach zero, achieving high μ_i (~1.0×10^4) and large ρ (~7.0×10^{-7} Ω·m) (Rassmann and Hofmann, 1968; Masumoto and Hinai, 1974).

10.1.3 Future Design of Crystalline Magnetic Alloys

Future work on the development of crystalline magnetic alloys may lie in two aspects, including compositional design and microstructural control. Possible principles for the compositional design can be summarized as: (i) enhanced exchange coupling for increased B_s with the addition of Co and/or Ni; (ii) reduced K_1 and λ_s for decreased H_c, for which elemental effect shown in Figure 10.1 (Hall, 1960) provides a suitable reference; and (iii) improved ρ resulted from alloying for reduced core loss.

Apart from the composition, microstructure also plays an important role in the magnetic performance. For instance, in the FeSi-based alloys, the Si may randomly substitute the Fe in the lattice to form the A2 disordered phase, B2 ordered phase with the Si occupying the body-centered location, as well as DO3 ordered phase with only half of the center sites occupied by the Si crossed diagonally (Figure 10.2). Despite many efforts devoted to identify the effects of the B2 and DO3 ordered phases on the soft

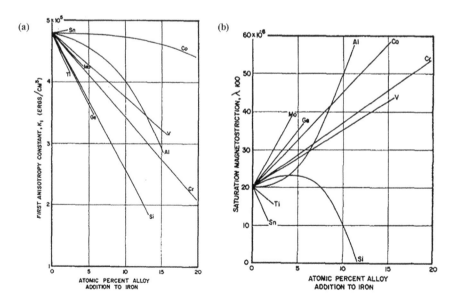

FIGURE 10.1 Influence of alloy elements on the anisotropy and magnetostriction of the Fe-based binary alloys. Reprinted with permission from (Hall, 1960). Copyright (1960) AIP Publishing.

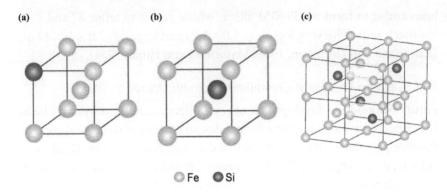

FIGURE 10.2 Schematic (a) A2, (b) B2 and (c) DO3 structures for the FeSi-based alloy.

magnetic performance, the findings are controversial, and the underlying mechanism remains unclear. In some investigations, the formation of B2 and DO3 ordered phases in the A2 matrix improves the soft magnetic properties (Narita and Teshima, 1984; Narita and Enokizono, 1979), while other studies indicate boundaries of B2 and DO3 phases hinder the motion of the domain walls for deteriorated H_c (Bouchara et al., 1990; Jung et al., 2011). Since the B2 and DO3 phases are metastable, finely controlled precipitation together with in-depth investigation on their growth mechanisms and relationship to the magnetic performance require further study.

10.2 AMORPHOUS MAGNETIC ALLOYS

The atoms exhibit long-range disorder for the amorphous magnetic alloys, which eliminates the magnetocrystalline anisotropy and crystal grain boundaries to impede the movement of magnetic domains. The amorphous magnetic alloys also exhibit relatively high ρ due to the scattering of electrons and the addition of non-metallic elements to enhance the glass-forming ability (GFA). Since Pol Duwez invented the rapid solidification method to produce amorphous AuSi alloy in 1960 (Klement et al., 1960) and subsequently developed the FePC amorphous magnetic alloy (Duwez and Lin, 1967), rapid growth of scientific and technological interest in the amorphous materials has been stimulated. Fe- and Co-based amorphous alloys have developed into commercial products, and the most widely used amorphous alloy system is the FeSiB due to its large B_s, high DC bias performance and relatively low cost (Kronmüller et al., 2007). The application of amorphous soft magnetic alloys was limited to winded toroid or stacked type due to the ribbon shape of alloys prepared by roller quenching method. The utilizing of amorphous powders had not been achieved commercially until the spinning water atomization process was developed in 1999 which made mass production of amorphous powders possible (Endo et al., 1999).

The design of amorphous alloys generally follows the empirical rules proposed by Akihisa. Inoue with the emphasis on three factors to achieve large GFA (Takeuchi and Inoue, 2005), including (i) material systems containing multiple elements (>3 types); (ii) distinctive difference (>12%) in the atomic size; and (iii) negative mixing enthalpy ΔH_{mix} between the constituent elements. Firstly, increased species of the elements are beneficial for disorderness and the GFA of the system (Xu et al., 2016). For instance, Nb, Mo and Cr have been incorporated into the $Fe_{72}Al_5Ga_2P_{11}C_6B_4$ which improves the supercooled liquid region width ΔT_x from 60 to 66 K (Makino et al., 2000). Secondly, the size of the constitutional atoms has also been varied for the formation of amorphous alloys (Lashgari et al., 2014; Fan and Shen, 2015). Small-sized atoms such as Si and P tend to occupy the space between large atoms, which is beneficial for forming more densely stacked structure for reduced free energy and stabilized supercooled liquid phase region (Shen et al., 2006; Xu et al., 2018). On the other hand, large-sized atoms such as Dy tend to enhance the atomic mismatch and form a skeleton reinforced structure for compactly stacked amorphousness (Li et al., 2018; Lin et al., 2010). Lastly, large negative ΔH_{mix} between the main elements of the alloy promotes the formation of short-range ordering in the liquid phase, which hinders atomic movement and delays the precipitation of crystal phase (Li et al., 2012; Xiao et al., 2020; Zhu et al., 2019). For example, Mn with large negative ΔH_{mix} (−45 kJ/mol for the Mn-Si, −32 kJ/mol for the Mn-B and −57.5 kJ/mol for the Mn-P) has been added into the FeSiBPCu alloy, giving rise to increased critical thickness for the formation of the amorphous alloy from 14 to 27 µm (Jia et al., 2018).

In summary, the amorphous alloys mainly consist of the ferromagnetic elements and those to improve the GFA. For the future design of high-performance amorphous magnetic alloys, the three empirical rules of Inoue still serve as important references. Alternatively, large GFA may also be achieved by increasing the cooling rate via special preparation routes. Either by composition design or fabrication innovation, enhanced GFA is beneficial to reduce the addition of the glass-forming elements, and thus to increase the content of the ferromagnetic elements. This is of great significance for enlarged B_s, which is crucial for high-power and miniaturization of related devices.

10.3 NANOCRYSTALLINE MAGNETIC ALLOYS

Nanocystalline magnetic alloys have been developed based on the amorphous systems, with unique microstructures of ultrafine grains embedded in the amorphous matrix. The randomly oriented and distributed nanograins give rise to both averaged K_1 and strong exchange coupling. Consequently, low H_c and large B_s can be achieved. To date, four main nanocrystalline systems including the Finemet, Nanoperm, Hitperm and Nanomet have been developed, which will be introduced in the following sections.

10.3.1 Finemet (FeCuNbSiB)

In 1988, Finemet with a unique microstructure of ultrafine α-Fe grains of 10–15 nm embedded in the amorphous matrix was firstly invented by Yoshizawa et al. by adding Cu and Nb into the FeSiB amorphous alloy (Yoshizawa et al., 1988). The structural evolution via melt-spinning followed by crystallization annealing is shown in Figure 10.3 (Hono et al., 1999). During the early annealing stage of the melt-spun amorphous ribbon, the Cu atoms of positive ΔH_{mix} with Fe tend to aggregate and form near-fcc clusters with a density of $10^{24}/m^3$ (Hono et al., 1999; Takeuchi and Inoue, 2005). The Cu clusters provide low energy interfaces upon which the activation energy for nucleation of the bcc α-Fe(Si) phase is reduced, giving rise to uniformly distributed Fe nuclei (Hono et al., 1999). Due to the hindrance effects of the large-radius Nb element on the diffusion of Fe atoms, the nuclei grow slowly and form an extremely fine nanocrystalline structure.

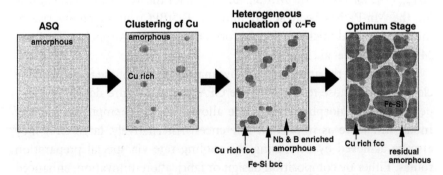

FIGURE 10.3 Schematic illustration of the crystallization process for the Finemet alloy. Reprinted with permission from (Hono et al., 1999). Copyright (1999) Elsevier.

The Finemet alloy with a composition of $Fe_{73.5}Cu_1Nb_3Si_{13.5}B_9$ possesses satisfactory soft magnetic performance of $\mu_e \sim 1 \times 10^5$, $H_c \sim 0.53$ A/m (Yoshizawa et al., 1988), which has been explained with the random anisotropy model (Herzer, 1990). Relatively large local anisotropy of the nanograins is averaged out across the basic ferromagnetic exchange length L_{ex}^0 described as

$$L_{ex}^0 = \sqrt{A/K_1} \tag{10.1}$$

where A is the exchange stiffness. For the $Fe_{80}Si_{20}$ with $A = 10^{-11}$ J/m and $K_1 = 8$ kJ/m³, the L_{ex}^0 is estimated as ~35 nm (Herzer, 1990). For grain size smaller than L_{ex}^0 in the Finemet, the ferromagnetic exchange interactions result in parallel alignment of the magnetic moments of the grains numbered at $N = (L_{ex}/D)^3$ where L_{ex} is the exchange length and D is the grain size. The averaged anisotropy <K> is given by Herzer (1990)

$$\langle K \rangle \approx \frac{K_1}{\sqrt{N}} = K_1 \left(\frac{D}{L_{ex}}\right)^{3/2} \tag{10.2}$$

According to Equation (10.1), the L_{ex} can be described as

$$L_{ex} = \sqrt{\frac{A}{\langle K \rangle}} \tag{10.3}$$

Combining Equation (10.2) and (10.3), we have

$$\langle K \rangle \approx \frac{K_1^4}{A^3} D^6 \tag{10.4}$$

Considering the results for coherent spin rotation, we get

$$H_c = p_c \frac{\langle K \rangle}{J_s} \approx p_c \frac{K_1^4 D^6}{J_s A^3} \tag{10.5}$$

$$\mu_i = p_\mu \frac{J_s^2}{\mu_0 \langle K \rangle} \approx p_\mu \frac{J_s^2 A^3}{\mu_0 K_1^4 D^6} \tag{10.6}$$

which is the famous D^{-6} law of nanocrystalline magnetic alloys. The p_c and p_μ are dimensionless constants, while J_s is the average saturation polarization. Consequently, nano-sized grain with a small D in the Finemet is beneficial for the averaged K_1, which gives rise to dramatically decreased H_c and increased μ_i.

To further tailor the magnetic properties of the Finemet, elemental substitution have been explored extensively, among which the majority of the studies have focused on the addition of ferromagnetic elements including Co (Wang et al., 2010; Mazaleyrat et al., 2004) and Ni (Agudo and Vázquez, 2005), as well as other transitional metals such as Mo, Ta, W and Ti (Yoshizawa and Yamauchi, 1991; Yan et al., 2010). In the $(Fe_{1-x}Co_x)_{73.5}Cu_1Nb_3Si_{13.5}B_9$ alloy system, the M_s reaches the maximum of 135 emu/g with $x = 0.2$, and the T_C of the amorphous phase increases until raised Co substitution of 50.0 at% (Mazaleyrat et al., 2004). However, the disadvantage of 50.0 at% Co substitution is the increased λ_s to 4.19×10^{-5} after annealing at 550 °C for 0.5 h, which is more than 10 times of that for the Co-free $Fe_{73.5}Cu_1Nb_3Si_{13.5}B_9$ (Wang et al., 2010). Replacing Fe with Ni in the Finemet gives rise to decreased T_C. With the Ni content surpassing 20.0 at%, tetragonal $Fe_3NiSi_{1.5}$ phase precipitates, which increases the H_c by up to two orders of magnitude (Agudo and Vázquez, 2005). For the $(Fe_{0.5}Co_{0.5})_{73.5-x}Ni_xSi_{13.5}B_9Nb_3Cu$, the initial permeability increases with Ni addition, whereas the T_C reaches its maximum at $x = 10$ followed by decreasing (Jia et al., 2011).

Effects of the addition of non-magnetic transitional elements have also been investigated. Elements including Ta, Mo, W, V and Cr have been selected to replace the Nb in the Finemet. They also exhibit hindrance effects on grain growth, following the sequence of Nb = Ta > Mo = W > V > Cr (Yoshizawa and Yamauchi, 1991). The substitution of Nb with Ti slightly increases the M_s of the Finemet due to raised volume fraction of the crystalline phase and the appearance of DO3 phase (Yan et al., 2010). The optimal Ti addition of 1 at% as in $Fe_{73.5}Cu_1Nb_2Ti_1Si_{13.5}B_9$ increases the temperature interval between the first and the second crystallization peaks, which promotes the precipitation of α-Fe(Si) phase over a wider temperature range.

10.3.2 Nanoperm (FeMBCu, M = Zr, Nb, Hf)

Although the Finemet exhibits excellent soft magnetic properties, its B_s is relatively small due to the limited content of Fe (70–83 at%). Based on the FeZr or FeHf amorphous system with higher Fe content (88–91 at%), new nanocrystalline alloy systems have been developed, which are commercially known as Nanoperm (FeMBCu, M = Zr, Nb, Hf). Typically, the FeZrBCu contains 5–7 at% Zr and 2–6 at% B, exhibiting large B_s of 1.6–1.7 T and relatively high μ_e of 1.4×10^4–2.0×10^4 (Suzuki et al., 2007). The alloy possesses microstructure with bcc nanocrystalline phase dispersed in the amorphous matrix enriched with Zr and B. The Zr clusters form according to the structural evolution of $Fe_{90}Zr_7B_3$ based on atom probe field ion microscopy, since Zr enrichment has been observed at the interfaces between the α-Fe grains and the amorphous matrix (Zhang et al., 1996). Minor B addition to the FeZr alloy increases the GFA, while excessive B content above 4.0 at% leads to inhomogeneity of grain size distribution (Kim et al., 1994).

It has been reported that up to 3.0 at% addition of Cu to $Fe_{87-x}Zr_7B_6Cu_x$ extends the temperature range for the precipitation of the crystalline phase and promotes homogeneous dispersion of the grains with finer grain size (Makino et al., 1991). Similar to the Finemet, the Cu atoms form clusters and act as preferential nucleation sites for the α-Fe via coherent interfaces (Ohkubo et al., 2001). The $Fe_{86}Zr_7B_6Cu_1$ alloy exhibits satisfactory soft magnetic properties of B_s = 1.52 T and $\mu_e = 4.8 \times 10^4$ as well as reduced core loss of 66 mW/kg (1T, 50Hz) (Suzuki et al., 1991).

10.3.3 Hitperm (FeCoMBCu, M = Zr, Nb, Hf)

High-temperature magnetic properties have been improved by adding Co into the Nanoperm in 1998, giving rise to a derivative nanocrystalline alloy system named as Hitperm (Willard et al., 1998). The resultant $(Fe_{0.5}Co_{0.5})_{88}Zr_7B_4Cu_1$ provides large B_s at elevated temperatures until transition from α to γ phase occurs at 980 °C. For the Hitperm, the Cu does not aggregate to form the nucleation sites for the α-Fe grain growth due to the smaller mixing enthalpy between Cu and Co than that between Cu and Fe (Ping et al., 2001).

10.3.4 Nanomet (FeSiBPCu)

Nanomet (FeSiBPCu), another nanocrystalline alloy system with high B_s of 1.8–1.9 T which is comparable to that of the FeSi, has attracted significant attention since 2009 (Makino et al., 2009a). Nanomet ribbons could be produced by melt-spinning in air atmosphere since it does not contain elements which are easily oxidized such as Zr and Nb (Makino et al.,

2003). Crystallization mechanism for the Nanomet is completely different from other nanocrystalline alloy systems discussed above, and exhibit a single amorphous phase in as-quenched state. For the Nanomet, high Fe content (>80 at%) yields the formation of primary α-Fe crystals in the as-quenched alloy (Makino et al., 2009b). Simultaneous addition of Cu (0.3–0.7 at%) and P (2.0–4.0 at%) into the $Fe_{82}Si_9B_9$ is beneficial to form a large amount of dispersive α-Fe nuclei (2–3 nm) in the amorphous matrix, due to the precipitation of Cu_3P-like clusters to provide lattice matching between the $(100)_{Cu3P}$ and $(111)_{Fe}$ for preferential nucleation of the α-Fe crystals (Makino, 2012).

Figure 10.4 illustrates the growth of the pre-existing and newly formed nanograins under different annealing conditions (Sharma et al., 2015). High heating rate (~ 400 °C/min) tends to promote the formation of extra nuclei above the new nucleation temperature T_{cn}. Such nuclei grow simultaneously with the pre-existing nuclei to generate a uniform microstructure. The competition between the growth of numerous α-Fe nuclei results in reduced grain size (< 20 nm). When the heating rate is low, there is sufficient time for the existing nuclei growth prior to the formation of any new nuclei, leading to a non-uniform structure. Specifically, at the temperature below T_{cn}, the annealing causes the formation of large grains based on the existing nuclei, while above T_{cn}, the pre-existing nuclei grow with the formation of new nuclei. With further raised temperature

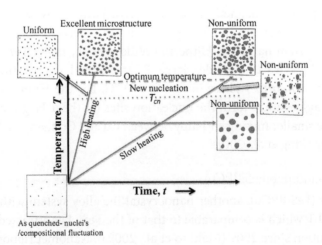

FIGURE 10.4 Crystallization mechanisms of the Nanomet. Reprinted with permission from (Sharma et al., 2015). Copyright (2015) Elsevier.

exceeding the optimum value, the pre-existing nuclei grow larger than the new formed nuclei, leading to a non-uniform nanocrystalline structure.

10.3.5 Future Design of Nanocrystalline Magnetic Alloys

During the evolution of nanocrystalline soft magnetic alloys, elements substitution plays a vital role in altering the microstructure, temperature characteristics and magnetic performance. The constituent elements of the nanocrystalline alloys can be summarized into four categories, including (i) Fe, Co and Ni as the ferromagnetic elements; (ii) elements with strong GFA such as Si, B and P; (iii) Cu to facilitate crystalline nucleation; and (iv) Nb, Zr, Hf and Ta with large atomic size to refrain further growth of the nanograins. To further improve the magnetic performance, increased content for the ferromagnetic element particularly the Fe, is desirable which should be balanced with the other properties such as GFA and brittleness. Also, high content of ferromagnetic elements tends to result in relatively high H_c due to the strong incoherent random anisotropy K_u (Suzuki and Herzer, 2012). Magnetic field-assisted annealing may be a future option for eliminating this effect to adjust the exchange coupling between different regimes of exchange coupled grains.

REFERENCES

Agudo, P., and M. Vázquez. 2005. Influence of Ni on the structural and magnetic properties of $Ni_xFe_{73.5-x}Si_{13.5}B_9Nb_3Cu_1$ ($0 \leq x \leq 25$) alloys. *Journal of Applied Physics* 97:023901.

Arnold, H. D., and G. W. Elmen. 1923. Permalloy, an alloy of remarkable magnetic properties. *Journal of the Franklin Institute* 195:621–32.

Bouchara, D., M. Fagot, and J. Degauque. 1990. Ordering Influence on magnetic properties of rapidly quenched Fe-6.5wt%Si. *Journal of Magnetism and Magnetic Materials* 83:377–8.

Bozorth, R. M. 1951. *Ferromagnetism*. New York: Van Nostrand.

Carr, W. J., and R. Smoluchowski. 1951. The magnetostriction of single crystals of iron-silicon alloys. *Physical Review* 83:1236–43.

Duwez, P., and S. C. H. Lin. 1967. Amorphous ferromagnetic phase in iron-carbon-phosphorus alloys. *Journal of Applied Physics* 38:4096–7.

Endo, I., I. Otsuka, R. Okuno, A. Shintani, M. Yoshino, and M. Yagi. 1999. Fe-based amorphous soft-magnetic powder produced by spinning water atomization process (SWAP). *IEEE Transactions on Magnetics* 35:3385–7.

Fan, X. D., and B. L. Shen. 2015. Crystallization behavior and magnetic properties in high Fe content FeBCSiCu alloy system. *Journal of Magnetism and Magnetic Materials* 385:277–81.

Fiorillo, F., G. Bertotti, C. Appino, and M. Pasquale. 1999. *Soft magnetic materials*. Philadelphia: John Wiley & Sons.

Gutfleisch, O., M. A. Willard, E. Bruck, C. H. Chen, S. G. Sankar, and J. P. Liu. 2011. Magnetic materials and devices for the 21st century: stronger, lighter, and more energy efficient. *Advanced Materials* 23:821–42.

Hall, R. C. 1959. Single crystal anisotropy and magnetostriction constants of several ferromagnetic materials including alloys of NiFe, SiFe, AlFe, CoNi, and CoFe. *Journal of Applied Physics* 30:816–9.

Hall, R. C. 1960. Single-crystal magnetic anisotropy and magnetostriction studies in iron-base alloys. *Journal of Applied Physics* 31:1037.

Hayakawa, M., K. Hayashi, W. Ishikawa, Y. Ochiai, H. Matsuda, Y. Iwasaki, and K. Aso. 1987. Soft magnetic properties of crystalline high B_s Fe-Co-Si and Fe-Co-Si-Al sputtered films. *IEEE Transactions on Magnetics* 23:3092–4.

Herzer, G. 1990. Grain size dependence of coercivity and permeability in nanocrystalline ferromagnets. *IEEE Transactions on Magnetics* 26:1397–402.

Hono, K., D. H. Ping, M. Ohnuma, and H. Onodera. 1999. Cu clustering and Si partitioning in the early crystallization stage of an $Fe_{73.5}Si_{13.5}B_9Nb_3Cu_1$ amorphous alloy. *Acta Materialia* 47:997–1006.

Jia, X. J., Y. H. Li, L. C. Wu, and W. Zhang. 2018. Structure and soft magnetic properties of Fe-Si-B-P-Cu nanocrystalline alloys with minor Mn addition. *AIP Advances* 8:056110.

Jia, Y. Y., Z. Wang, R. M. Shi, J. Yang, H. J. Kang, and T. Lin. 2011. Influence of Ni addition on structure and magnetic properties of FeCo-based Finemet-type alloys. *Journal of Applied Physics* 109:073917.

Jung, H. J., S. B. Kim, J. B. Kim, and J. R. Kim. 2011. Effects of anti-phase boundary on the Iron loss of grain-oriented silicon steel. *ISIJ International* 51:987–90.

Kim, K. Y., T. H. Noh, and I. K. Kang. 1994. Microstructural change upon annealing Fe-Zr-B alloys with different boron contents *Materials Science and Engineering A* 179–180:552–6.

Klement, W., R. H. Willens, and P. Duwez. 1960. Non-crystalline structure in solidified gold-silicon alloys. *Nature* 187:869–70.

Kronmüller, H., S. Parkin, R. Waser, U. Böttger, and S. Tiedke. 2007. *Handbook of magnetism and advanced magnetic materials*. Hoboken: John Wiley & Sons.

Lashgari, H. R., D. Chu, S. S. Xie, H. D. Sun, M. Ferry, and S. Li. 2014. Composition dependence of the microstructure and soft magnetic properties of Fe-based amorphous/nanocrystalline alloys: a review study. *Journal of Non-Crystalline Solids* 391:61–82.

Li, W. C., C. X. Xie, Y. Z. Yang, and C. L. Yao. 2018. Glass formation and soft magnetic properties in Dy-containing Fe-Si-B alloys by adjusting B/Si mole ratio. *Journal of Non-Crystalline Solids* 489:1–5.

Li, X., Y. Zhang, H. Kato, A. Makino, and A. Inoue. 2012. The effect of Co addition on glassy forming ability and soft magnetic properties of Fe-Si-B-P bulk metallic glass. *Key Engineering Materials* 508:112–6.

Lin, S. L., S. F. Chen, J. K. Chen, and Y. L. Lin. 2010. Formation and magnetic properties of Fe-Si-B-Dy amorphous alloy. *Intermetallics* 18:1826–28.

Makino, A. 2012. Nanocrystalline soft magnetic Fe-Si-B-P-Cu alloys with high of 1.8–1.9 T contributable to energy saving. *IEEE Transactions on Magnetics* 48:1331–5.

Makino, A., T. Bitoh, A. Inoue, and T. Masumoto. 2003. Nb-Poor Fe-Nb-B nanocrystalline soft magnetic alloys with small amount of P and Cu prepared by melt-spinning in air. *Scripta Materialia* 48:869–74.

Makino, A., A. Inoue, and T. Mizushima. 2000. Soft magnetic properties of Fe-based bulk amorphous alloys. *Materials Transactions, JIM* 41:1471–7.

Makino, A., H. Men, T. Kubota, K. Yubuta, and A. Inoue. 2009a. FeSiBPCu nanocrystalline soft magnetic alloys with high B_s of 1.9 tesla produced by crystallizing hetero-amorphous phase. *Materials Transactions* 50:204–9.

Makino, A., H. Men, K. Yubuta, and T. Kubota. 2009b. Soft magnetic FeSiBPCu heteroamorphous alloys with high Fe content. *Journal of Applied Physics* 105:013922.

Makino, A., K. Suzuki, A. Inoue, and T. Masumoto. 1991. Low core loss of a bcc $Fe_{86}Zr_7B_6Cu_1$ alloy with nanoscale grain size. *Materials Transactions, JIM* 32:551–6.

Masumoto, H., and M. Hinai. 1974. Magnetic properties of high permeability alloys "Hardperm" in the Ni-Fe-Nb system. *Transactions of the Japan Institute of Metals* 38:261–4.

Masumoto, H., and T. Yamamoto. 1937. On a new alloy "Sendust" and its magnetic and electric properties Fe-Si-Al. *Transactions of the Japan Institute of Metals* 1:127–35.

Mazaleyrat, F., Z. Gercsi, J. Ferenc, T. Kulik, and L. K. Varga. 2004. Magnetic properties at elevated temperatures of Co substituted Finemet alloys. *Materials Science and Engineering A* 375–377:1110–5.

Narita, K., and M. Enokizon. 1978. On magnetic properties of Fe-Si-Mn alloy sheet. *IEEE Transactions on Magnetics* 14:365–7.

Narita, K., and M. Enokizono. 1979. Effect of ordering on magnetic properties of 6.5-percent silicon-iron alloy. *IEEE Transactions on Magnetics* 15:911–5.

Narita, K., and N. Teshima. 1984. Magnetic properties of rapidly quenched silicone-iron ribbons. *Journal of Magnetism and Magnetic Materials* 41:86–92.

Ohkubo, T., H. Kai, D. H. Ping, K. Hono, and Y. Hirotsu. 2001. Mechanism of heterogeneous nucleation of α-Fe nanocrystals from $Fe_{89}Zr_7B_3Cu_1$ amorphous alloy. *Scripta Materialia* 44:971–6.

Ouyang, G. Y., X. Chen, Y. F. Liang, C. Macziewski, and J. Cui. 2019. Review of Fe-6.5 wt%Si high silicon steel—a promising soft magnetic material for sub-kHz application. *Journal of Magnetism and Magnetic Materials* 481:234–50.

Ping, D. H., Y. Q. Wu, and K. Hono. 2001. Microstructural characterization of $(Fe_{0.5}Co_{0.5})_{88}Zr_7B_4Cu_1$ nanocrystalline alloys. *Scripta Materialia* 45:781–6.

Rassmann, G., and U. Hofmann. 1968. Classification of high-permeability nickel-iron alloys. *Journal of Applied Physics* 39:603–5.

Sato-Turtelli, R., A. Penton-Madrigal, C. F. Barbatti, R. Grössinger, H. Sassik, E. Estevez-Rams, R. S. Sarthour, E. H. C. P. Sinnecker, and A. P. Guimarães. 2005. Effect of the addition of Cr, Ta and Nb on structural and magnetic properties of Fe-Si alloys. *Journal of Magnetism and Magnetic Materials* 294:e151-4.

Sharma, P., X. Zhang, Y. Zhang, and A. Makino. 2015. Competition driven nanocrystallization in high B_s and low coreloss Fe-Si-B-P-Cu soft magnetic alloys. *Scripta Materialia* 95:3-6.

Shen, B. L., M. Akiba, and A. Inoue. 2006. Effects of Si and Mo additions on glass-forming in FeGaPCB bulk glassy alloys with high saturation magnetization. *Physical Review B* 73:104204.

Sourmail, T. 2005. Near equiatomic FeCo alloys: constitution, mechanical and magnetic properties. *Progress in Materials Science* 50:816-80.

Suzuki, K., and G. Herzer. 2012. Magnetic-field-induced anisotropies and exchange softening in Fe-rich nanocrystalline soft magnetic alloys. *Scripta Materialia* 67:548-53.

Suzuki, K., N. Kataoka, A. Inoue, A. Makino, and T. Masumoto. 2007. High saturation magnetization and soft magnetic properties of bcc Fe-Zr-B alloys with ultrafine grain structure. *Materials Transactions, JIM* 31:743-6.

Suzuki, K., A. Makino, A. Inoue, and T. Masumoto. 1991. Soft magnetic properties of nanocrystalline bcc Fe-Zr-B and Fe-M-B-Cu (M = Transition metal) alloys with high saturation magnetization. *Journal of Applied Physics* 70:6232.

Takeuchi, A., and A. Inoue. 2005. Classification of bulk metallic glasses by atomic size difference, heat of mixing and period of constituent elements and its application to characterization of the main alloying element. *Materials Transactions* 46:2817-29.

Wakiyama, T., M. Takahashi, S. Nishimaki, and J. Shimoda. 1981. Magnetic properties of Fe-Si-Al single crystals. *IEEE Transactions on Magnetics* 17:3147-50.

Wang, Z., J. Yang, Y. M. Han, Z. D. Xhang, B. Fu, and R. C. Ye. 2010. Magnetostriction and effective magnetic anisotropy of Co-contained Finemet nanocrystalline alloys. *Journal of Applied Physics* 107:09A308.

Willard, M. A., D. E. Laughlin, M. E. McHenry, D. Thoma, K. Sickafus, J. O. Cross, and V. G. Harris. 1998. Structure and magnetic properties of $(Fe_{0.5}Co_{0.5})_{88}Zr_7B_4Cu_1$ nanocrystalline alloys. *Journal of Applied Physics* 84:6773.

Xiao, M., Y. Z. Yang, Z. G. Zheng, X. Liu, Q. Zhou, Z. G. Qiu, and D. C. Zeng. 2020. Comparative study of V and Mo on the glass formation and magnetic properties of Fe-Si-B-Cu-Nb-M (M = V, Mo) nanocrystalline for high frequency applications. *Journal of Non-Crystalline Solids* 546:120294.

Xu, J., Y. Z. Yang, W. Li, X. C. Chen, and Z. W. Xie. 2016. Effect of P addition on glass forming ability and soft magnetic properties of melt-spun FeSiBCuC alloy ribbons. *Journal of Magnetism and Magnetic Materials* 417:291-3.

Xu, J., Y. Z. Yang, W. C. Li, Z. W. Xie, and X. C. Chen. 2018. Effect of Si addition on crystallization behavior, thermal ability and magnetic properties in high Fe content Fe-Si-B-P-Cu-C alloy. *Materials Research Bulletin* 97:452–456.

Yan, M., H. Tong, S. Tao, and J. H. Liu. 2010. Structural and magnetic properties of $Fe_{73.5}Cu_1Nb_{3-x}Ti_xSi_{13.5}B_9$ ($x \leq 3$) alloys. *Journal of Alloys and Compounds* 505:264–7.

Yoshizawa, Y., S. Oguma, and K. Yamauchi. 1988. New Fe-based soft magnetic alloys composed of ultrafine grain structure. *Journal of Applied Physics* 64:6044.

Yoshizawa, Y., and K. Yamauchi. 1991. Magnetic properties of Fe-Cu-M-Si-B (M = Cr, V, Mo, Nb, Ta, W) alloys. *Materials Science and Engineering A* 133:176–9.

Zhang, Y., K. Hono, A. Inoue, A. Makino, and T. Sakurai. 1996. Nanocrystalline structural evolution in $Fe_{90}Zr_7B_3$ soft magnetic material. *Acta Materialia* 44:1497–510.

Zhu, M., M. Zhang, L. J. Yao, R. H. Nan, Z. Y. Jian, and F. Chang. 2019. Effect of Mo substitution for Nb on the glass-forming ability, magnetic properties, and electrical resistivity in $Fe_{80}(Nb_{1-x}Mo_x)_5B_{15}$ ($x = 0$–0.75) amorphous ribbons. *Vacuum* 163:368–72.

FURTHER READING

Talaat, A., M. V. Suraj, K. Byerly, A. Wang, Y. Wang, J. K. Lee, and P. R. Ohodnicki Jr. 2021. Review on soft magnetic metal and inorganic oxide nanocomposites for power applications, *Journal of Alloys and Compounds* 870:159500.

CHAPTER 11

Soft Magnetic Composites

Soft magnetic composites (SMCs) have been developed based on the magnetic alloys. As shown in Figure 11.1 (Strečková et al., 2014), by wrapping the metallic powder with insulation coating, followed by compaction into the magnetic core, it is possible to maintain large magnetization and high permeability of the SMCs since the magnetic metallic alloy remains as the dominant component. Due to the separation of the metallic powders with the insulation coatings for significantly reduced eddy current loss, the SMCs are critical for high-frequency applications. The compaction, on one hand, allows fabrication of SMCs with flexible shapes for different application occasions. On the other hand, it generates internal stress which should be eliminated by subsequent annealing.

Research on SMCs has been developed along (i) the design of material systems based on crystalline, amorphous and nanocrystalline soft magnetic alloys via elemental addition or substitution, and (ii) optimizing processing technologies including powder production, insulation coating, compaction and annealing. Since developments of the various metallic magnetic systems have been described in Chapter 10, recent progress on the fabrication technologies of SMCs will be provided in the following sections.

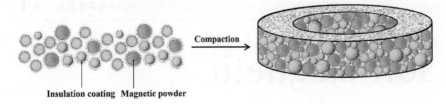

FIGURE 11.1 Schematic illustration showing the fabrication of soft magnetic composite. Reprinted with permission from (Strečková et al., 2014) Copyright (2014) Elsevier.

11.1 POWDER PRODUCTION AND SIZE DISTRIBUTION

The fabrication of SMCs starts from the preparation of metallic magnetic powders, mainly via mechanical crushing or atomization. Both routes involve initial melting the raw materials with designed elemental ratio. Powers made from mechanical crushing exhibit irregular shape. Although such shape is beneficial to achieve satisfactory stacking density, the edges and corners of the powders may damage the insulation coating during compaction, resulting in increased core loss. Both gas atomization and water atomization have been used in powder fabrication as well. Powders made via gas atomization possess spherical shape and low oxygen content, while those made by water atomization exhibit irregular shape with higher oxygen content. For the powders with spherical shape, uniform and complete coating can be obtained with higher tolerance to compaction. Compared with the powders of irregular shape, spherical powders however, may lead to the more gaps in-between the powders for reduced permeability and saturation magnetization of the SMCs.

Not only should the effects of the powder shape on the performance of the SMCs be considered, but also appropriate methods for powders preparation should be chosen according to the material systems. For the malleable Fe, FeNi and FeNiMo alloys, atomization is usually used to prepare the corresponding powders, whereas the brittle FeSi and FeSiAl alloys can not only be easily broken into small pieces followed by ball-milling into powders, but also be atomized for powder preparation. In order to achieve amorphous and nanocrystalline powders, fast solidification is essential for which atomization and melt-spinning are commonly used. The atomization method involves using high-speed fluid or air to impact alloy melts into small droplets with a high cooling rate to form amorphous powders as shown in Figure 11.2a (Zhao et al., 2021). Generally, water atomization

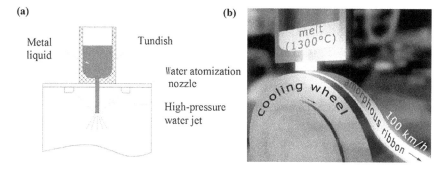

FIGURE 11.2 Schematics of (a) water atomization and (b) melt-spinning. (a) Reprinted with permission from (Zhao et al., 2021). Copyright (2021) Elsevier. (b) Reprinted with permission from (Herzer, 2013). Copyright (2013) Elsevier.

is more favorable to prepare amorphous powders due to the large cooling rate. The melt-spinning method has also been utilized to prepare amorphous ribbons where the alloy melts at around 1300 °C and falls onto a high-speed rotating copper roller with a linear velocity of ~100 km/h. This provides a fast cooling rate of 10^5–10^6 °C/s to form the amorphous ribbons with thickness of 20–30 μm, as shown in Figure 11.2b (Herzer, 2013). The ribbons can then be directly wounded into a core for applications in transformers or crushed into powders to fabricate SMCs.

Size distribution of the powders also exerts significant impact on the performance of the SMCs. Generally, the SMCs containing large-sized particles exhibit larger magnetization and permeability due to the relatively higher content of the magnetic component, whereas small-sized particles with a larger combined surface area may involve better insulation coatings for reduced eddy current loss. In practice, mixed particle sizes are used, allowing the small particles to fill in the gaps between large particles to achieved higher density of SMCs with comprehensively enhanced magnetic properties and decreased loss.

11.2 INSULATION COATING

Insulation coating is critical to prepare SMCs with enlarged electrical resistivity and lowered loss. Depending on the coating material, it can be divided into organic coatings, inorganic coatings and their combination.

11.2.1 Organic Coatings

The most commonly used organic coatings include epoxy resin (Hemmati et al., 2006; Shokrollahi et al., 2009), phenolic resin (Taghvaei et al., 2009a) and silicone resin (Wu et al., 2012b). Effects of the resin content, usually between 1.0 wt% and 5.0 wt%, on the magnetic performance of the SMCs have been investigated (Chicinas et al., 2007; Shokrollahi et al., 2009). On one hand, increasing the coating content results in enlarged electrical resistivity and reduced eddy current loss as well as increased mechanical strength. On the other hand, the introduction of non-magnetic resin leads to reduced magnetization and permeability.

Considering the hydrophilic nature of the magnetic powders which may be incompatible with the organic insulation, silane coupling agent has been employed as modifier to improve the uniformity and adhesion coating between magnetic particles and the phenolic resin (Taghvaei et al., 2009a). Such modification results in enhanced wetting between the metal powder and the wrapping resin, which is beneficial to achieve lower loss compared with the SMCs without the coupling agent.

Since compaction tends to induce internal stress which deteriorates the coercivity, subsequent annealing is usually necessary for stress relaxation. Consequently, endeavors have been made to improve the thermal stability of the resin coatings. Poor thermal stability of the epoxy and phenolic resin usually limits the annealing temperature used (< 200 °C) which is insufficient for complete stress relieving. Wu et al. have utilized silicone resin with better thermal stability as the coating layer for the Fe powders (Wu et al., 2012b). According to differential thermal analysis, the annealing temperature that the silicone resin could stand in air reaches ~600 °C. Compared to the SMCs with conventional epoxy resin coating, those containing the silicone resin exhibit reduced loss over a wide frequency range after annealing at 580 °C.

Although the silicone resin provides improved thermal stability, it exhibits inferior strength to the epoxy resin. Consequently, organic-silicone epoxy resin (OER) has been used in the fabrication of Fe SMCs, combining the advantages of silicone resin and epoxy resin with satisfactory thermal stability and mechanical properties (Xiao et al., 2013). For a given coating content and annealing temperature, the bond strength between the Fe powders and the OER may be further increased via vacuum annealing. Wu et al. have utilized the polymer parylene C as the insulating coating in the fabrication of Fe SMCs (Wu et al., 2015), which exhibits lower friction factor, high molecular weight and better crystallinity for enhanced resistivity to chemical attack and higher density stability (increment of 17.02% compared with the conventional epoxy resin).

11.2.2 Inorganic Coatings

Inorganic coatings can be roughly divided into two categories depending on whether the alloy component in the magnetic powders is involved in the coating reaction. For direct reactions with the alloy components, acids have usually been used, including non-oxidizing acids like phosphorous acid (Taghvaei et al., 2009b) or chromic acid (Zhang et al., 2014), and oxidizing acid such as nitric acid (Kang et al., 2007; Nakahara et al., 2010). Phosphate coating is used both experimentally and industrially due to its satisfactory adhesiveness to the magnetic powders (Taghvaei et al., 2009b). Depending on the alloy composition, phosphorization may give rise to varied resultants as discussed by Huang et al. (2015b). Oxidizing acids tend to generate hydroxides which then evolve into oxides after annealing (Kang et al., 2007; Nakahara et al., 2010). Liu et al. have studied the influences of HNO_3 concentration and reaction time on the properties of FeSiAl SMCs, with the volution of the coatings demonstrated in Figure 11.3 (Liu et al., 2018). With 10 wt% HNO_3, the oxidation process represents steady-state passivation while the increased HNO_3 concentration of 30 wt% results in pitting corrosion during the oxidation. Apart from the acids, passivation

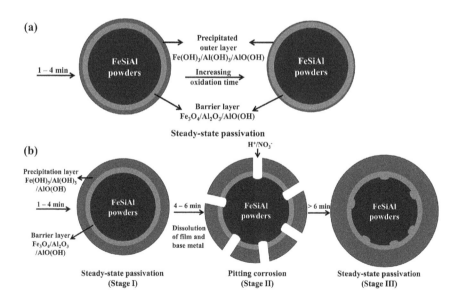

FIGURE 11.3 Schematics showing the evolution of the coating layer for the FeSiAl powders oxidized with (a) 10 wt% and (b) 30 wt% HNO_3. Reprinted with permission from (Liu et al., 2018). Copyright (2018) Elsevier.

by oxidizing salts, e.g. NaNO$_2$ (Taghvaei et al., 2010b), have also been investigated with the impacts of passivation conditions such as pH of the solution and reaction time revealed. Generally, these parameters affect the reaction rate and Gibbs energy level during the coating, which determine the amount and species of the resultants, i.e. the thickness and composition of the coatings.

To avoid magnetic dilution by non-magnetic coatings, ferrimagnetic materials such as Fe$_3$O$_4$, MnZn and NiZn ferrites have been employed as the insulation coatings. Zhao et al. have compared the magnetic properties of Fe SMCs fabricated via surface oxidation by H$_2$O and O$_2$ with those fabricated via traditional phosphate coating (Zhao et al., 2016). The SMCs with the resultant Fe$_3$O$_4$ coatings exhibit larger magnetization and effective permeability together with lower core loss compared with the phosphate-coated ones, as shown in Figure 11.4 (Zhao et al., 2016). Fe$_3$O$_4$

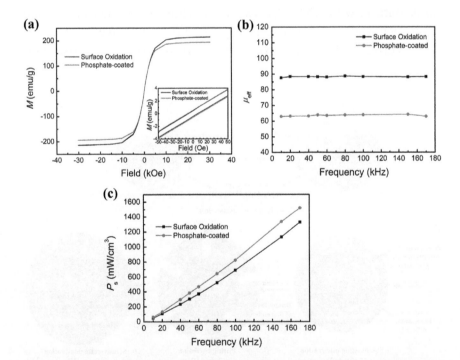

FIGURE 11.4 (a) Hysteresis loops for the surface oxidized and phosphate-coated Fe powders, (b) effective permeability and (c) core loss as a function of frequency for the Fe SMCs prepared by surface oxidation and phosphate coating. Reprinted with permission from (Zhao et al., 2016). Copyright (2016) Elsevier.

insulation for the Fe SMCs may also be achieved via other methods such as alkaline and acidic bluing (Zhao et al., 2015). Apart from the Fe_3O_4, MnZn ferrites (Wu et al., 2012a) and NiZn ferrites (Guo et al., 2020) have also been fabricated via sol-gel and solid-state reaction to provide simultaneously maintained magnetization and low loss.

The other category of inorganic coatings does not involve reaction with the magnetic powders, providing generality to be applied to various material systems. For instance, sol-gel method has been used to produce fine and homogeneous oxide insulation coatings such as Al_2O_3 (Peng et al., 2016), MgO (Taghvaei et al., 2010a) and SiO_2 (Pang et al., 2006) with high resistivity and thermal stability. Factors such as the precursor, pH and reaction time have been taken into consideration to achieve suitable coatings, since thin coatings may be insufficient to reduce the intra-particle eddy current loss while thick coatings not only tend to crack during compaction, but also induce magnetic dilution and increased coercivity. Wu et al. have employed hydrolysis precipitation of $Al(NO_3)_3$ in the growth of Al_2O_3 coatings for FeSiAl SMCs and revealed two growth mechanisms with varied pH conditions (Wu et al., 2018a). When pH = 3, massive H^+ reacts with the Fe and Al prior to the Al^{3+} hydrolysis. Raised pH of 8 gives rise to direction reaction between the Al^{3+} with the OH^- in the formation of the $Al(OH)_3$ colloids on the FeSiAl powder surfaces.

The above strategies of direct reaction and non-involvement of the powder element may be combined. Liu et al. have developed a double-layered phosphate-alumina coating via initial direction reaction of the FeSiAl powders with the phosphate acid followed by sol-gel formation of the Al_2O_3 (Liu et al., 2015). The hybrid coating provides superior overall magnetic performance of the SMCs compared with sole phosphate or Al_2O_3 coating, since the inner phosphate layer provides strong adhesion and the outer Al_2O_3 layer hinders the decomposition of the phosphate.

11.2.3 Hybrid Organic–Inorganic Coatings

In recent years, there has been a trend of introducing inorganic materials into the organic polymer resin, forming hybrid inorganic-organic insulating coating to improve both magnetic and mechanical properties of the SMCs (Strečková et al., 2015; Pang et al., 2007). For instance, inorganic additives including SiO_2 and $ZnSO_4$ have been introduced into phenolic resin to suppress the water release during curing and to enhance the thermal stability of the hybrid resin (Strečková et al., 2012). Strečková et al.

have adopted the sol-gel method for in situ growth of nano-silica particles in the phenol-formaldehyde resin (PFR) modified with 3-glycidoxypropyltrim (Strečková et al., 2013). Fourier transform infrared spectroscopy (FTIR) analysis indicates chemical bonding between the phenolic resin and the silica, while thermal gravimetric (TG) and differential scanning calorimeter (DSC) analysis demonstrate decomposition temperature of the hybrid around 70 °C higher than that of the original PFR. The SMCs prepared with the hybrid insulation coating exhibit increased permeability and reduced eddy current loss. In a later study, the same group has compared hybrid organic–inorganic phenolic resins via in situ growth of silica and boron nanoparticles respectively (Strečková et al., 2015). It has been shown that incorporation of the boron into the phenolic resin gives rise to enhanced mechanical strength by nearly three times compared with the SMCs with silica-modified phenolic resin.

Besides the phenolic resins, silicone resin has also been incorporated with the nano-silica particles, exhibiting increased electrical resistivity and magnetic permeability with raised content of the organic phase (Pang et al., 2007). Wu et al. have introduced SiO_2 nanoparticles into epoxy-modified silicone resin (ESR) via in situ sol-gel growth (Wu et al., 2018b). FTIR reveals chemical bonds between the SiO_2 and the epoxy, ester and hydroxyl groups of the ESR as illustrated in Figure 11.5a, which is beneficial for improved permeability and reduced loss. Furthermore, both Fe_3O_4 and SiO_2 have been introduced into the ESR for optimized magnetic performance (Figure 11.5b) (Luo et al., 2018). While the addition of Fe_3O_4 reduces magnetic dilution, the in situ growth of the SiO_2 facilitates the dispersion of the Fe_3O_4 along with further increasing the electrical resistivity of the hybrid.

11.3 COMPACTION

Compaction of the magnetic powders is necessary to produce SMCs with desired shapes. On one hand, higher compaction pressure gives rise to increased density of the composites for larger magnetization and permeability. On the other hand, excessive pressure may result in damaged insulation coatings as well as increased stress for the magnetic powders with deteriorated coercivity. The optimal compaction conditions may vary for different material systems. For the Fe SMCs, a pressure of 800 MPa is usually sufficient since the Fe is relatively soft and can be easily deformed (Taghvaei et al., 2010a, 2009a). Contrastingly, the FeSi (Zhou et al., 2019)

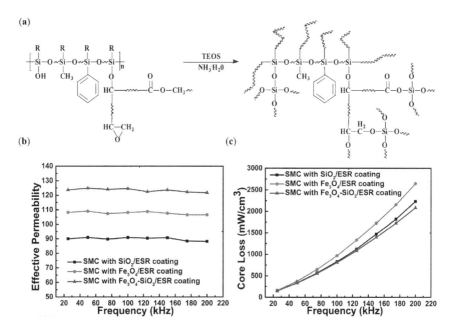

FIGURE 11.5 (a) Reaction to form the chemical bonding of the silanol with the hydroxyl, epoxy and ester groups of the ESR. Reprinted with permission from (Wu et al., 2018b). Copyright (2018) Elsevier. Variations of the (b) effective permeability and (c) core loss with the frequency for the SMCs containing the SiO_2/ESR, Fe_3O_4/ESR and Fe_3O_4-SiO_2/ESR coatings. Reprinted with permission from (Luo et al., 2018). Copyright (2018) Elsevier.

and FeSiAl (Liu et al., 2018) SMCs require higher pressure of around 2000 MPa. The compaction pressure for FeNi SMCs is similar with the Fe SMCs (Koohkan et al., 2008; Olekšáková et al., 2013) while the compaction pressure for FeNiMo SMCs reaches up to 1800 MPa (Yao et al., 2020; Zhang et al., 2014). For the amorphous and nanocrystalline alloys such as FeSiB (Kim et al., 2007; Wang et al., 2020) and FeCuNbSiB (Huang et al., 2015a; Moon et al., 2007), compaction pressures between 1800–2100 MPa are necessary.

11.4 ANNEALING

Effects of annealing are also two-folds. It is beneficial to eliminate the stress induced by compaction while it may also damage the insulation coating with poor thermal stability for increased eddy current loss. Consequently, appropriate annealing temperature should be chosen taking both factors

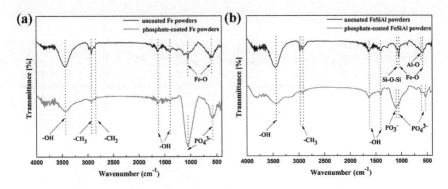

FIGURE 11.6 FTIR spectra of phosphate-coated (a) Fe and (b) FeSiAl powders before and after annealing. Reprinted with permission from (Huang et al., 2015b). Copyright (2015) Elsevier.

into consideration. For the Fe SMCs with relatively mild compaction, an annealing temperature of 500 °C is sufficient to eliminate the internal stress, while for the FeSi and FeSiAl systems, temperatures between 700–800 °C are necessary. Jang et al. have investigated the effects of annealing temperature on the properties of the FeSi SMCs and reported decreased hysteresis loss but increased eddy current loss with raised temperature, and the minimum loss can be achieved at 800 °C (Jang et al., 2008). Huang et al. have studied the evolution of phosphate coating during annealing for Fe and FeSiAl SMCs, and found that the optimal temperature for Fe and FeSiAl were 500 °C and 700 °C respectively (Huang et al., 2015b). As revealed by the FTIR spectra in Figure 11.6, when annealed at high temperature (≥600 °C), the $FePO_4$ coating formed surrounding the Fe powders tends to decompose into iron phosphide with deteriorated electrical resistivity while the $Al(PO_3)_3$ coating on the surfaces of the FeSiAl powders converts into Al_2O_3 and P_2O_5 which continues to function as the insulation layers.

11.5 CHALLENGES AND PERSPECTIVES

One key challenge in front of the research on SMCs is to reduce the magnetic loss with minimum magnetic dilution. In this regard, developing new insulator coatings with simultaneous high electrical resistivity and soft magnetic property is promising. The formation of uniform and thin coatings with satisfactory adhesion to the magnetic powders is also desirable. On the other hand, compaction and annealing are involved during the fabrication, and their optimized conditions may vary for different material systems. With the possible development of new metallic magnetic materials, it is crucial to investigate the most suitable fabrication conditions for the individual system.

REFERENCES

Chicinas, I., O. Geoffroy, O. Isnard, and V. Pop. 2007. AC magnetic properties of the soft magnetic composites based on nanocrystalline Ni-Fe powders obtained by mechanical alloying. *Journal of Magnetism and Magnetic Materials* 310:2474–6.

Guo, R. D., S. M. Wang, Y. Zhong, K. Sun, X. N. Jiang, G. H. Wu, C. J. Wu, and Z. W. Lan. 2020. FeSiCr@NiZn SMCs with ultra-low core losses, high resistivity for high frequency applications. *Journal of Alloys and Compounds* 830:154736.

Hemmati, I., H. R. Madaah Hosseini, and A. Kianvash. 2006. The correlations between processing parameters and magnetic properties of an iron-resin soft magnetic composite. *Journal of Magnetism and Magnetic Materials* 305:147–51.

Herzer, G. 2013. Modern soft magnets: Amorphous and nanocrystalline materials. *Acta Materialia* 61:718–34.

Huang, C. B., T. C. Liu, X. Y. Wang, C. W. Lu, D. R. Li, and Z. C. Lu. 2015a. Magnetic properties of nanocrystalline powder cores fabricated by mechanically crushed powders. *Journal of Iron and Steel Research International* 22:67–71.

Huang, M. Q., C. Wu, Y. Z. Jiang, and M. Yan. 2015b. Evolution of phosphate coatings during high-temperature annealing and its influence on the Fe and FeSiAl soft magnetic composites. *Journal of Alloys and Compounds* 644:124–30.

Jang, P. W., B. H. Lee, and G. Choi. 2008. Effects of annealing on the magnetic properties of Fe-6.5%Si alloy powder cores. *Journal of Applied Physics* 103:07E743.

Kang, E. Y., Y. H. Chung, M. R. Ok, and H. K. Baik. 2007. Research on the surface oxidation procedure of Fe-base metallic glass during wet oxidation treatment. *Materials Science and Engineering A* 449–451:159–64.

Kim, Y. B., D. H. Jang, H. K. Seok, and K. Y. Kim. 2007. Fabrication of Fe-Si-B based amorphous powder cores by cold pressing and their magnetic properties. *Materials Science and Engineering A* 449–451:389–93.

Koohkan, R., S. Sharafi, H. Shokrollahi, and K. Janghorban. 2008. Preparation of nanocrystalline Fe-Ni powders by mechanical alloying used in soft magnetic composites. *Journal of Magnetism and Magnetic Materials* 320:1089–94.

Liu, D., C. Wu, and M. Yan. 2015. Investigation on sol-gel Al_2O_3 and hybrid phosphate-alumina insulation coatings for FeSiAl soft magnetic composites. *Journal of Materials Science* 50:6559–66.

Liu, D., C. Wu, M. Yan, and J. Wang. 2018. Correlating the microstructure, growth mechanism and magnetic properties of FeSiAl soft magnetic composites fabricated via HNO_3 oxidation. *Acta Materialia* 146:294–303.

Luo, D. H., C. Wu, and M. Yan. 2018. Incorporation of the Fe_3O_4 and SiO_2 nanoparticles in epoxy-modified silicone resin as the coating for soft magnetic composites with enhanced performance. *Journal of Magnetism and Magnetic Materials* 452:5–9.

Moon, B. G., K. Y. Sohn, W. W. Park, and T. D. Lee. 2007. Effect of milling on the soft magnetic behavior of nanocrystalline alloy cores. *Materials Science and Engineering A* 449:426–9.

Nakahara, S., E. A. Périgo, Y. Pittini-Yamada, Y. De Hazan, and T. Graule. 2010. Electric insulation of a FeSiBC soft magnetic amorphous powder by a wet chemical method: Identification of the oxide layer and its thickness control. *Acta Materialia* 58:5695–703.

Olekšáková, D., J. Füzer, P. Kollár, and S. Roth. 2013. Components of the core losses under low frequency magnetic field of the bulk Ni-Fe compacted powder material. *Journal of Magnetism and Magnetic Materials* 333:18–21.

Pang, Y. X., S. N. B. Hodgson, J. Koniarek, and B. Weglinski. 2007. The influence of the dielectric on the properties of dielectromagnetic soft magnetic composites. Investigations with silica and silica hybrid sol-gel derived model dielectric. *Journal of Magnetism and Magnetic Materials* 310:83–91.

Pang, Y. X., S. N. B. Hodgson, B. Weglinski, and D. Gaworska. 2006. Investigations into sol-gel silica and silica hybrid coatings for dielectromagnetic soft magnetic composite applications. *Journal of Materials Science* 41:5926–36.

Peng, Y. D., Y. Yi, L. Y. Li, J. H. Yi, J. W. Nie, and C. X. Bao. 2016. Iron-based soft magnetic composites with Al_2O_3 insulation coating produced using sol-gel method. *Materials & Design* 109:390–5.

Shokrollahi, H., K. Janghorban, F. Mazaleyrat, M. L. Bue, V. Ji, and A. Tcharkhtchi. 2009. Investigation of magnetic properties, residual stress and densification in compacted iron powder specimens coated with polyepoxy. *Materials Chemistry and Physics* 114:588–94.

Strečková, M., R. Bureš, M. Fáberová, L. Medvecký, J. Füzer, and P. Kollár. 2015. A comparison of soft magnetic composites designed from different ferromagnetic powders and phenolic resins. *Chinese Journal of Chemical Engineering* 23:736–43.

Strečková, M., J. Füzer, L. Kobera, J. Brus, M. Fáberová, R. Bureš, P. Kollár, M. Lauda, Ľ. Medvecký, V. Girman, H. Hadraba, M. Baťková, and I. Baťko. 2014. A comprehensive study of soft magnetic materials based on FeSi spheres and polymeric resin modified by silica nanorods. *Materials Chemistry and Physics* 147:649–60.

Strečková, M., Ľ. Medvecký, J. Füzer, P. Kollár, R. Bureš, and M. Fáberová. 2013. Design of novel soft magnetic composites based on Fe/resin modified with silica. *Materials Letters* 101:37–40.

Strečková, M., T. Sopčák, Ľ. Medvecký, R. Bureš, M. Fáberová, I. Batko, and J. Briančin. 2012. Preparation, chemical and mechanical properties of micro-composite materials based on Fe powder and phenol-formaldehyde resin. *Chemical Engineering Journal* 180:343–53.

Taghvaei, A. H., A. Ebrahimi, K. Gheisari, and K. Janghorban. 2010a. Analysis of the magnetic losses in iron-based soft magnetic composites with MgO insulation produced by sol-gel method. *Journal of Magnetism and Magnetic Materials* 322:3748–54.

Taghvaei, A. H., H. Shokrollahi, A. Ebrahimi, and K. Janghorban. 2009a. Soft magnetic composites of iron-phenolic and the influence of silane coupling agent on the magnetic properties. *Materials Chemistry and Physics* 116:247–53.

Taghvaei, A. H., H. Shokrollahi, and K. Janghorban. 2009b. Magnetic and structural properties of iron phosphate–phenolic soft magnetic composites. *Journal of Magnetism and Magnetic Materials* 321:3926–32.

Taghvaei, A. H., H. Shokrollahi, and K. Janghorban. 2010b. Structural studies, magnetic properties and loss separation in iron-phenolicsilane soft magnetic composites. *Materials & Design* 31:142–8.

Wang, C., Z. L. Guo, J. Wang, H. B. Sun, D. C. Chen, W. H. Chen, and X. Liu. 2020. Industry-oriented Fe-based amorphous soft magnetic composites with SiO_2-coated layer by one-pot high-efficient synthesis method. *Journal of Magnetism and Magnetic Materials* 509:166924.

Wu, C., X. W. Gao, G. L. Zhao, Y. Z. Jiang, and M. Yan. 2018a. Two growth mechanisms in one-step fabrication of the oxide matrix for FeSiAl soft magnetic composites. *Journal of Magnetism and Magnetic Materials* 452:114–9.

Wu, C., M. Q. Huang, D. H. Luo, Y. Z. Jiang, and M. Yan. 2018b. SiO_2 nanoparticles enhanced silicone resin as the matrix for Fe soft magnetic composites with improved magnetic, mechanical and thermal properties. *Journal of Alloys and Compounds* 741:35–43.

Wu, S., A. Z. Sun, Z. W. Lu, and C. Cheng. 2015. Fabrication and properties of iron-based soft magnetic composites coated with parylene via chemical vapor deposition polymerization. *Materials Chemistry and Physics* 153:359–64.

Wu, S., A. Z. Sun, W. H. Xu, Q. Zhang, F. Q. Zhai, P. Logan, and A. A. Volinsky. 2012a. Iron-based soft magnetic composites with Mn-Zn ferrite nanoparticles coating obtained by sol-gel method. *Journal of Magnetism and Magnetic Materials* 324:3899–905.

Wu, S., A. Z. Sun, F. Q. Zhai, J. Wang, Q. Zhang, W. H. Xu, P. Logan, and A. A. Volinsky. 2012b. Annealing effects on magnetic properties of silicone-coated iron-based soft magnetic composites. *Journal of Magnetism and Magnetic Materials* 324:818–22.

Xiao, L., Y. Sun, C. Ding, L. Yang, and L. Yu. 2013. Annealing effects on magnetic properties and strength of organic-silicon epoxy resin-coated soft magnetic composites. *Proceedings of the Institution of Mechanical Engineers, Part C: Journal of Mechanical Engineering Science* 228:2049–58.

Yao, Z. X., Y. D. Peng, C. Xia, X. W. Yi, S. H. Mao, and M. T. Zhang. 2020. The effect of calcination temperature on microstructure and properties of FeNiMo@ Al_2O_3 soft magnetic composites prepared by sol-gel method. *Journal of Alloys and Compounds* 827:154345.

Zhang, Z. M., W. Xu, T. Guo, Y. Z. Jiang, and M. Yan. 2014. Effect of processing parameters on the magnetic properties and microstructures of molybdenum permalloy compacts made by powder metallurgy. *Journal of Alloys and Compounds* 594:153–7.

Zhao, G. L., C. Wu, and M. Yan. 2015. Fe-based soft magnetic composites with high B_s and low core loss by acidic bluing coating. *IEEE Transactions on Magnetics* 51:1–4.

Zhao, G. L., C. Wu, and M. Yan. 2016. Enhanced magnetic properties of Fe soft magnetic composites by surface oxidation. *Journal of Magnetism and Magnetic Materials* 399:51–7.

Zhao, T. C., C. G. Chen, X. J. Wu, C. Z. Zhang, A. A. Volinsky, and J. J. Hao. 2021. FeSiBCrC amorphous magnetic powder fabricated by gas-water combined atomization. *Journal of Alloys and Compounds* 857:157991.

Zhou, M. M., Y. Han, W. W. Guan, S. J. Han, Q. S. Meng, T. T. Xu, H. L. Su, X. Guo, Z. Q. Zou, F. Y. Yang, and Y. W. Du. 2019. Magnetic properties and loss mechanism of Fe-6.5wt%Si powder core insulated with magnetic Mn-Zn ferrite nanoparticles. *Journal of Magnetism and Magnetic Materials* 482:148–54.

FURTHER READING

Périgo, E. A., B. Weidenfeller, P. Kollár, and J. Füzer, 2018. Past, present, and future of soft magnetic composites, *Applied Physics Reviews* 5:031301.

CHAPTER **12**

Soft Magnetic Ferrites

12.1 BASICS OF SOFT MAGNETIC FERRITES

Magnetic materials are required to function over a wide range of frequencies from the static direct current (DC) to even GHz (Pardavi-Horvath, 2000). Since the eddy current loss increases sharply in proportion to the square of the frequency, the family of Fe-based metallic materials with small electrical resistivity become unsuitable for high frequencies. Alternatively, soft ferrites as ferrimagnetic transition-metal oxides with electrically insulating characteristics have been developed in the early 1930s, and soon become an important category of magnetic materials in commercial production. Compared to Fe and FeSi with electrical resistivity ρ in the range of 10^{-7}–10^{-5} $\Omega \cdot$m, soft ferrites possess higher ρ of 10–10^6 $\Omega \cdot$m (Goldman, 2006), giving rise to remarkable reduction in the eddy current. The eddy current loss becomes negligible, as a decisive advantage of ferrites at high frequencies with less heat generation. For the electrical steel FeSi, the value of skin depth d decreases from 0.36 mm at 50 Hz to 3.6 μm at 500 kHz, thus the assembling lamination thickness should be less than d to permit the magnetic flux penetration and to minimize core loss (Coey, 2010), while soft ferrites are free of such problems. Ferrites are also advantageous due to their unique combination of excellent chemical stability, low material cost and near-net shape processing (van der Zaag, 2021). It is also necessary to point out that due to the ferrimagnetic nature

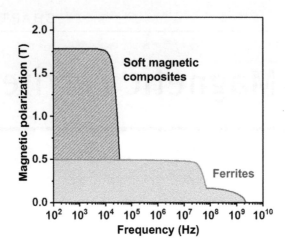

FIGURE 12.1 Comparison of the magnetic polarization between the soft magnetic composites and ferrites in different frequencies.

and the non-magnetic oxygen atoms, ferrites exhibit mediocre polarization of 0.2–0.5 T, which is far inferior to that of the metallic ferromagnets (Pullar, 2012; Vedrtnam et al., 2020).

Figure 12.1 illustrates the preferred materials for different frequency ranges. Generally, the soft magnetic composites are intended for applications below 100 kHz range (Shokrollahi and Janghorban, 2007), above which, typically at the MHz range, soft ferrites with inherently large ρ and reasonable magnetic properties become the materials of choice. Statistics show that soft ferrites are produced at a quantity of ~0.3 million tons per year, representing about 25% of the global market for all the soft magnetic materials (Coey, 2010). Nowadays, China has evolved into the geographical center for large-scale ferrite production, accounting for ~85% of the worldwide total production.

12.2 CRYSTAL STRUCTURE OF SOFT FERRITES

The crystal structure of soft ferrite is spinel (cubic symmetry, space group $Fd3m$) with the general chemical formula of $MO \cdot Fe_2O_3$ or MFe_2O_4. Here M represents divalent metallic ions including Fe^{2+}, Co^{2+}, Ni^{2+}, Mn^{2+}, Zn^{2+}, or Mg^{2+}, or more often, a combination of these ions (Kharisov et al., 2019). As introduced in Chapter 1, the lodestone or the magnetite $FeO \cdot Fe_2O_3$ is the earliest and naturally occurred ferrite, which has been applied in the navigation compasses more than two millennia ago. M may also refer to two types of ions with different valences (one with M^{3+}, the other one with M^{1+}), or possibly the mixture of trivalent ions and vacant sites. Consequently, it provides diverse possibilities to formulate the mixed ferrites with delicately tailored properties catering to desired applications.

As schematically shown in Figure 12.2, two types of interstices coexist between the oxygen anions, namely the tetrahedrally coordinated site (indicated by A), and the octahedrally coordinated site (indicated by B). In the unit cell containing 8 formula units of MFe_2O_4 (56 atoms including 32 oxygen anions and 24 metal cations), there are 64 tetrahedral A sites and 32 octahedral B sites to accommodate the metallic M and Fe. However, to fulfill the ionic charge balance, only 8 of the A sites and 16 of the B sites are occupied by cations. This gives rise to many possible cationic substitutions, given that the crystal structure remains thermodynamically stable during solid state reaction.

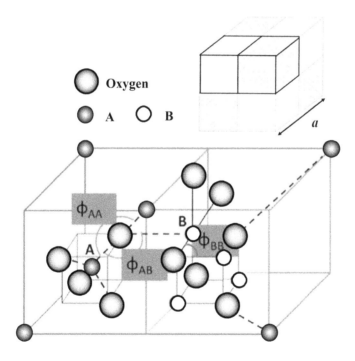

FIGURE 12.2 Schematic of the spinel structure. The unit cell contains 8 subcells, two of which are displayed. The large spheres represent the oxygen anions to form the tetrahedral A site and octahedral B site. Super-exchange interaction between A and B sublattices are also displayed.

The ultimate magnetic properties strongly depend on the cation distribution over the A/B sites. Referred as the normal spinel structure, all M^{2+} cations accommodate at the tetrahedral A site and Fe^{3+} cations occupy the octahedral B site (Soliman et al., 2011; Ehrhardt et al., 2003; Burdett et al., 1982). Examples of normal spinel include $ZnO \cdot Fe_2O_3$ and $CdO \cdot Fe_2O_3$. For the inverse spinel, the trivalent Fe^{3+} ions are equally divided between A/B sites, while the divalent M^{2+} ions are accommodated at B sites (Szotek et al., 2006). Since the M^{2+} cations with larger radius than that of the trivalent one prefers to enter the larger B site, it becomes reasonable that most of the soft ferrites belong to the inverse spinel structure. The ions distributions can be explicitly written as $\left[M^{2+}_\delta Fe^{3+}_{1-\delta}\right]_A \left[M^{2+}_{1-\delta} Fe^{3+}_{1+\delta}\right]_B O^{2-}_4$, with $\delta = 0$ for the inverse spinel ferrite, $\delta = 1$ for the normal spinel ferrite, and $0 < \delta < 1$ for the mixed spinel ferrite (Mugutkar et al., 2020).

There is a wide variety of ferrites with different compositions and remarkably varied magnetic behaviors. The site preference of different ions is fundamental to compare the magnetic properties quantitatively, which is determined by the joint effects from the ionic radii, the interstices size, temperature and valence of the cations. Among the three types of super-exchange interactions, j_{A-B}, j_{B-B}, and j_{A-A} via the intermediate O^{2-} (Figure 12.2), the antiferromagnetic A–B interaction usually dominates with the smaller bond length and relatively large bond angle. Comparably, the B–B interaction with larger bond length is much weaker, and A–A interaction with a bond angle of ~80° is the weakest (Nakamura et al., 2003). Table 12.1 shows the structural ion distribution and calculated net moment for a series of basic ferrites from $MnFe_2O_4$ to $ZnFe_2O_4$ (Kotnala and Shah, 2015). For the inverse spinel ferrite, the moment of Fe^{3+} ions evenly distributed at the A and B sites is offset, so that the saturation magnetization M_s comes from the net moment of the divalent M ions at the B site. This is demonstrated in Figure 12.3, showing the magnetic structure of Fe_3O_4 as a typical example. From the intrinsic magnetic properties given in Table 12.2, it is evident that the Curie temperature T_C of different ferrites ranges from 575 to 943 K, and saturation magnetization M_s from 0.18 MA/m to 0.50 MA/m (Coey, 2010). Almost all of them possess the [111] magnetically easy direction, low magnetocrystalline anisotropy constant K_1, and low to moderate magnetostriction λ_s.

On the basis of the simple ferrites such as $MnFe_2O_4$ and $NiFe_2O_4$, different solid solution strategies are used to prepare the commercial ferrites

TABLE 12.1 Metallic Ionic Distribution and Magnetic Moment of Simple Ferrites (Kotnala and Shah, 2015)

Ferrites	Ionic Distribution		Magnetic Moment of Tetrahedral Ion (μ_B)	Magnetic Moment of Octahedral Ion (μ_B)	Magnetic Moment per Molecule (μ_B)	
	Tetrahedral Site	Octahedral Site			Theoretical	Experimental
$MnFe_2O_4$	Fe^{3+}	$Fe^{3+} + Mn^{2+}$	5	5 + 5	5	4.6
Fe_3O_4	Fe^{3+}	$Fe^{2+} + Fe^{3+}$	5	5 + 4	4	4.1
$CoFe_2O_4$	Fe^{3+}	$Fe^{3+} + Co^{2+}$	5	5 + 3	3	3.7
$NiFe_2O_4$	Fe^{3+}	$Fe^{3+} + Ni^{2+}$	5	5 + 2	2	2.3
$CuFe_2O_4$	Fe^{3+}	$Fe^{3+} + Cu^{2+}$	5	5 + 1	1	1.3
$MgFe_2O_4$	Fe^{3+}	$Fe^{3+} + Mg^{2+}$	5	5 + 0	0	1.1
$Li_{0.5}Fe_{2.5}O_4$	Fe^{3+}	$Fe^{3+}_{1.5} + Li^{+}_{0.5}$	5	7.5 + 0	2.5	2.6
$CdFe_2O_4$	Cd^{2+}	$Fe^{3+} + Fe^{3+}$	0	5-5	0	1
$ZnFe_2O_4$	Zn^{2+}	$Fe^{3+} + Fe^{3+}$	0	5-5	0	1

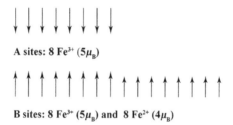

A sites: 8 Fe^{3+} (5μ_B)

B sites: 8 Fe^{3+} (5μ_B) and 8 Fe^{2+} (4μ_B)

FIGURE 12.3 Ferrimagnetic ordering of Fe_3O_4, for which ions at the A and B sites are antiferromagnetically coupled (antiparallel), while the neighboring ions at the B sites are ferromagnetically coupled (parallel).

depending on their ultimate application requirements. Currently, all the commercial ferrites are mixed ferrites (Costa et al., 2003; Thakur et al., 2020). Two broad classes with massive production worldwide are (Mn, Zn) O·Fe_2O_3 (MnZn) ferrite and (Ni, Zn)O·Fe_2O_3 (NiZn) ferrite, as first predicted by Snoek (Snoek, 1949). Introduced non-magnetic Zn^{2+} ions usually occupy the A sites and reduce the A-sublattice magnetization to increase the net magnetic moment of the MnZn and NiZn soft ferrites (Smit and Wijn, 1959). MnZn ferrites with an initial permeability of 1000–5000 (as high as 40,000 for high-permeability MnZn), coercivity of < 1 Oe and electrical resistivity of 1–100 Ω·m are suitable for operational frequency up to 3 MHz (Goldman, 2006). Comparably, NiZn ferrites are intended for

TABLE 12.2 Room Temperature Magnetic Properties of Simple Ferrites (Coey, 2010)

Ferrites	a_0 (pm)	T_C (K)	M_s (MA/m)	K_1 (kJ/m³)	λ_s (10⁻⁶)	ρ (Ω·m)
$MnFe_2O_4$	852	575	0.50	−3	−5	10^5
Fe_3O_4	840	860	0.48	−13	40	10^{-1}
$CoFe_2O_4$	839	790	0.45	290	−110	10^5
$NiFe_2O_4$	834	865	0.33	−7	−25	10^2
$MgFe_2O_4$	836	713	0.18	−3	−6	10^5
$Li_{0.5}Fe_{2.5}O_4$	829	943	0.33	−8	−8	1

1–300 MHz or higher frequency band due to even higher resistivity (>10^3 Ω·m), but with lower polarization of ~0.3 T, initial permeability of 10–1000 and coercivity of several Oe (Coey, 2010).

12.3 POWER LOSS OF SOFT FERRITES

For soft ferrites, crucial magnetic properties include saturation induction, permeability, power loss and their temperature dependence. Since the operational frequency has increased from kHz to MHz and even higher ranges for the power supplies, there arises an urgent need to reduce the power loss of soft ferrites. As introduced in Chapter 9, the total power loss P_c is the sum of hysteresis loss P_h, eddy current loss P_e and residual loss P_r, which can be written as (van der Zaag, 1999)

$$P_c = P_h + P_e + P_r = C_H B^3 f + C_E B^2 f^2 / \rho + P_r \qquad (12.1)$$

where C_H and C_E are dimensional constants of hysteresis loss and eddy current loss, B and ρ refer to the magnetic flux density and electrical resistivity of a given material, and f is frequency. Via plotting the P_c/f curve versus f (Figure 12.4), both P_h and P_e can be separated, with the remaining dissipation as P_r.

The P_h generated by the irreversible domain wall movement is proportional to the area of hysteresis loop. The hysteresis loss can then be reduced upon a narrow S-shaped loop. Consequently, an appropriate composition with nera-zero K_1 and λ_s is desired to facilitate the domain wall movement and avoid the hysteresis loss (Ishino and Narumiya, 1987). From the microstructural perspective, uniform grain growth, low porosity, low level of impurities and stresses can also minimize the P_h.

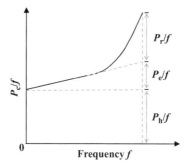

FIGURE 12.4 Schematic diagram of the total power loss P_c versus the frequency f for soft ferrites, with the three component losses scaled by the frequency.

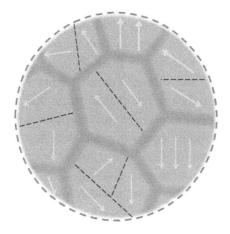

FIGURE 12.5 Schematic illustration showing the microstructure of the polycrystalline ferrite.

The P_e is heavily dependent on the electrical resistivity of the magnetic material. Therefore soft ferrites with higher electrical resistivity are required at higher operational frequencies. Since the ceramic ferrite is polycrystalline consisting of crystal grains and grain boundaries (GBs) (Figure 12.5), the most common approaches to suppress eddy current loss include increasing the resistivity of the grain boundary or the grains. For the former, CaO-SiO$_2$ co-dopants with minimal solubility into the lattice are introduced, which tend to diffuse toward the GBs and form the CaSiO$_3$ glassy phase as an insulating layer. Other co-dopants to separate the grains include CaO-ZrO$_2$, SnO$_2$-TiO$_2$ etc. For the latter, since the electrical conductivity within the grains is mainly attributed to the electron hopping between the trivalent Fe^{3+} and divalent Fe^{2+} cations, retaining the Fe^{3+} state at the octahedral site becomes a necessity. Introducing multivalent ions (Ti^{4+}, Sn^{4+} and so on) into the spinel lattice is an effective strategy, which pair with the excessive Fe^{2+}, reducing the electron hopping for increased grain resistivity. Sometimes, even Fe-deficient composition is designed for high-frequency applications. More details on the composition design of the soft ferrites such as Fe/Zn/Co concentration and multi-dopants can be referred to the book recommended in the Further Reading section at the end of this chapter.

The P_r is significant for the dissipation reduction up to MHz range. It has been found that the P_r increases at a fast rate and may account for over 80% of the P_c under frequencies above 500 kHz (Otobe et al., 1999). Microscopic mechanisms controlling the P_r lie in the ferromagnetic resonance and intragranular domain wall resonance, both of which can be eliminated in fine-grained ferrites through using finer powders or low-temperature sintering for shorter periods (Shokrollahi, 2008). For the mono-domain grains (below the critical grain size of 3–4 μm for the MnZn ferrite) (van der Zaag, 1999), intragranular domain walls become absent. Low-melting-point sintering aided with Bi_2O_3 is commonly utilized to yield small grains (Gore et al., 2015). Grain growth inhibitor is also necessary to control the gain size. The Co^{3+} doping can freeze the domain wall movement, serving as a widely adopted strategy to suppress the P_r at the sacrifice of permeability and hysteresis loss. However, at higher frequencies, the origin of the power loss gradually changes from domain wall movement to magnetic domain rotation. On such occasions, the magnetic loss is almost independent of the specific microstructure of the material, but is only related to its composition (Beatrice et al., 2008).

12.4 APPLICATIONS OF SOFT FERRITES

It is a difficult task to satisfy all the magnetic requirements simultaneously. In most cases, the choice of different ferrites is a trade-off between polarization, permeability, power losses and cost. In the following, we will briefly review the main features of the commercially available ferrites for different applications.

12.4.1 High Frequency Ferrites

Under static and small magnetic field, the initial permeability μ_i of metals can reach up to 10^5, while the soft ferrites exhibit the maximum μ_i of approximately 10^4. When applying an alternating magnetic field of increased frequency, the μ_i of metals tends to decrease significantly. Distinctively, the μ_i of ferrites almost maintains over a wide frequency range until a critical roll-off frequency f_r. Figure 12.6 compares the complex permeability as a function of frequency for typical MnZn and NiZn ferrites. Apparently, the higher the static value of μ_i, the lower the roll-off frequency, following the Snoek's relation. High frequency and high permeability are mutually incompatible. Consequently, if constant inductance is required for the ferrite core, which indicates constant permeability at all frequencies up to hundreds of MHz, ferrites with low permeability and high roll-off frequency are necessary.

12.4.2 High Permeability Ferrites

For applications in telecommunications with a weak AC magnetic field, such as the antenna coils and signal transmission transformers, the applied field H is so small that the ferrite cannot be magnetized to saturation. In such occasions, the soft ferrites are utilized to efficiently sense and securely amplify the small input signals, so large permeability is an important parameter on the crystal structure, microstructure, chemical composition, temperature, stress and so on. The crystalline anisotropy and magnetostriction that determine the movement of domains and affect the permeability should be minimized (Shokrollahi, 2008). From

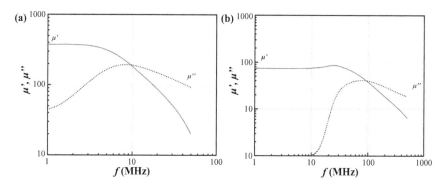

FIGURE 12.6 Complex permeability as a function of frequency for typical (a) MnZn and (b) NiZn ferrites.

Table 12.2, only Fe_3O_4 possesses a positive magnetostriction. Hence to obtain high μ_i, slightly more than 50 mol% Fe_2O_3 should be designed to compensate the negative magnetostriction values of other ferrites. The addition of ZnO is beneficial to increase the magnetic moment, lower the anisotropy and magnetostriction for enhanced permeability. Since Zn is rather volatile, the Zn loss during sintering should be considered. Permeability is highly sensitive to the impurities in the raw material, which may cause ionic size mismatch, subsequent distortion and strain of the lattice. For example, the content of SiO_2 should be controlled, since excessive SiO_2 leads to a duplex structure of large grains within a fine grain for reduced permeability (Ott et al., 2003). Consequently, to prepare high-quality high-permeability ferrites, the purity of raw material should be sufficient. Furthermore, higher permeability can be achieved with larger grain size and decreased porosity. Comparably, the porosity at GB regions is less detrimental to the permeability of ferrites than that within the grains.

After continual technological developments, the permeability of commercial ferrites has been greatly increased. Taking the available TDK ferrites as typical examples, the initial permeability increases from ~3300 for the H5A ferrite, to 7500 for the H5B2, 10000 for the H5C2, 15000 for the H5C3, further to 30000 for the H5C5 ferrite. A typical permeability versus temperature curve of MnZn ferrite is shown in Figure 12.7. The right peak drops to zero near the Curie point while the left peak refers to the secondary maximum of permeability (SMP), which generally occurs at the point with the magnetostriction approaching zero. The SMP can be tuned by proper choice of the composition, which is usually designed around room temperature for low-end devices, and 60–100 °C for power ferrites.

12.4.3 Power Ferrites

Power ferrites are intended for power transformers and the like, which convert high-frequency pulse current to the required voltage and are called as switching AC adapters. With strong AC magnetic field applied, power ferrites should possess large μ_i and B_s, as well as low P_c. These switching power supplies are applied in most electronic devices using commercial alternating current such as TVs, DVD recorders and gaming consoles. The netlike high-resistance

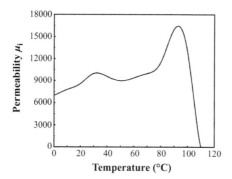

FIGURE 12.7 Plot of permeability versus temperature of typical MnZn ferrite.

GBs, where the additives prefer to segregate, plays a key role in the low-loss polycrystal ferrites. Co-doping of CaO and SiO_2 is commonly used to reduce the power loss. Substitution strategy with tetravalent and pentavalent oxide such as TiO_2, SnO_2, V_2O_5 and Ta_2O_5 has also been employed to combine with the Fe^{2+} and suppress the electron hopping with the contribution of Fe^{2+} maintained to compensate the magnetostriction (Ishino and Narumiya, 1987). Yan et al. have reported that multi-dopants (including La, Ti, Si and Ca) for GB engineering of MnZn ferrite give rise to the optimized P_c of 267 kW/m³ (5 MHz, 10 mT, 100 °C) and μ_i of 644, superior to the previously reported results (Yan et al., 2021). It has also been found that large crystallographic mis-orientations than 25° changes the domain structure from multi- to mono-domain state, offering an alternative way to design high-frequency and low-loss soft ferrites.

Power loss of the ferrites is considerably reduced yearly due to accumulated scientific understanding and advanced manufacturing technologies. The working frequency for the MnZn and NiZn ferrites has reached 3 MHz and 10 MHz, respectively. For power ferrites, the power loss in the unit of mW/cm³ are often measured at the intended service temperatures higher than the room temperature. The representative ferrite manufacturers include TDK, FDK, Ferroxcube, DMEGC and TDG, which supply commercial ferrites under various trade names. For instance, the main trademarks of MnZn ferrites for 3 MHz applications include 7H20 (FDK from Japan), PC200 (TDK from Japan), 3F4 (Ferroxcube from the

Netherlands), DMR50B (DMEGC from China) and TP5E (TDG from China).

12.5 MANUFACTURING TECHNOLOGY OF SOFT FERRITES

Thus far, we have discussed the important properties of soft ferrites for specific applications. Ferrites are widely recognized as the structure-sensitive materials. Besides the chemical composition design, the desired performance of ferrites depends on the microstructural engineering during the manufacturing process. In this section, we will discuss the processing technology of the soft ferrites. Unlike the soft magnetic composites described in Chapter 11, the soft ferrites are usually manufactured via the ceramic route with the following procedures.

(1) Raw material selection on the basis of economic consideration and desired magnetic properties. During processing, a common goal for ferrites is to form the spinel structure. Based on the simple spinel-type ferrites, judicious combination of simple ferrite at appropriate ratios induces a wide variety of commercial ferrites. For the MnZn ferrites, the main component consists of Fe_2O_3, ZnO and MnO in the powder form, the ratios of which depend on different applications. Usually, the base Fe concentration is designed to be higher than the stoichiometric 50 mol% (Vangroenou et al. 1969). For the NiZn ferrites, the main component consists of (50–70) mol% Fe_2O_3, (5–40) mol% ZnO, and (5–40) mol% NiO.

(2) Grinding. The powder mixtures after prolonged grinding by ball-milling exhibit a smaller particle size, which can decrease the porosity of the ultimate products. Subsequently, the introduced water during wet grinding process should be removed in a filter press, and the ferrites are loosely pressed and dried.

(3) Pre-sintering. The mixed oxides are calcined at ~1000 °C by batch or rotary calciner. Solid state reaction takes place between the constituents where the elements inter-diffuse to form partial ferrites under the driving force of concentration gradient. The calcining process is accompanied with certain shrinkage, which reduces the shrinkage during the final sintering and allows uniform final product.

(4) Milling and granulation. Since the powders coarsen considerably after calcining, further grinding is needed to mix the remaining unreacted oxides and lower the particle size by ball-milling or attritors (generally in the order of 1 μm or smaller). A small proportion of organic binder is mixed together and spray-dried to form granules suitable for the following pressing.

(5) Pressing or extrusion. The granules are poured into the die and compressed to prepare the green compact. Most shapes including toroidal cores can be achieved by pressing while for long rods and tubes, extrusion is used.

(6) Sintering. As a critical step closely related to the magnetic properties of the ferrites, the sintering cycle differs with varied compositions and desired properties. The basic requirements include complete inter-diffusion between adjacent particles, formation of spinel structure and reduced porosity. To increase the density, the sintering agents, including polyvinyl alcohol, CaO, SiO_2 and Bi_2O_3, can be added beforehand. For NiZn ferrites, the furnace temperature and duration time should be strictly controlled to balance the high density, grain growth and zinc loss. Given the fact that lower loss can be achieved with inhibited domain wall motion at high frequencies, sometimes the NiZn soft ferrites are intentionally underfired with resultant porosity interference with the wall motion, which contribute to lowering the μ_i and P_c. Unlike NiZn ferrite that the Ni^{2+} remains essentially stable, the sintering atmosphere control is crucial for high-performance MnZn ferrite (Sugimoto, 1999), which is rather complex with the existence of Mn^{2+}, Mn^{3+}, Zn^{2+}, Fe^{2+}, Fe^{3+} and O^{2-}. In an ideal furnace atmosphere, the partial pressure of oxygen shall change with different temperatures to be consistent with the equilibrium oxygen pressure of the soft ferrite.

12.6 FUTURE PERSPECTIVES OF SOFT FERRITES

It is undoubtable that soft ferrites are indispensable components in the electronic devices in the future. The advent of 5G era puts forward higher requirements on the electronic technologies and interrelated magnetic materials. The following aspects are expected to attract continuous attention from the scientists and engineers.

Firstly, in-depth basic research needs to be conducted to solve the long-standing obstacles in the domain tuning of the soft ferrites. The main difficulty is the power loss mechanism, particularly for the residual loss, which limits the high-frequency applications. Also, precise control of the oxygen partial pressure equilibrium, together with the oxidation and reduction kinetics of the ferrites are challenging, especially for the MnZn ferrite. Moreover, the synergetic effects of multi-component dopants on the magnetic and electric properties of soft ferrites remain a fundamental research subject (Yan et al., 2021). The progress of accurate characterizations, including extended x-ray absorptance fine structure (EXAFS), Mössbauer spectroscopy, three-dimensional atom probe (3DAP) and Lorentz transmission electron microscopic (L-TEM) analysis contributes to analyzing the ionic distribution and the hyperfine structure of ferrites (Liu et al., 2021a). Meanwhile, the introduction of computational material science can efficiently simulate the microstructural evolution, magnetization process and forecast the magnetic properties, which provide important theoretical basis for developing high-performance ferrites (Fiorillo et al., 2014).

Secondly, novel ferrites catering to new applications are in urgent need. Presently, growing applications in 13.56 MHz near-field communication, wireless charging and 5G base station require new apparatus that enables efficient mass production of ultra-thin and flexible ferrite sheets at low cost, high quality and reliability. Inserting the ferrite sheet between metal surface and antenna prevents the reaction of magnetic flux on the metal surface of battery pack. It is extremely important for the continual progress of ferrite chip inductor industry. Its flexibility allows application with high processability, where customized shapes, thickness and sizes are available upon request. Besides, it is difficult to achieve simultaneous low-loss and high-permeability in the entire frequency range. Developing a composite material of ferrite and Fe-based metallic alloy may be prospective since the disadvantage of low saturation magnetization of ferrite may be compensated, and its advantage of high electrical resistivity can be maintained (Birčáková et al., 2020). Nanostructured ferrites with single-domain nanocrystals below the critical size are also emerging to avoid the domain wall resonance for higher working frequencies (Shirsath et al.,

2018). Numerous efforts have been devoted to develop nanocrystalline ferrites by sol-gel (Manju and Raji, 2018), hydrothermal synthesis (Phumying et al., 2013, Chen et al., 2018) and mechanical alloying (Jalaly et al., 2009). Based on these advances, novel manufacture techniques have been developed, including two-step sintering (Sun et al., 2021; Yang and Wang, 2016), spark plasma sintering (Jenus et al., 2021) and three-dimensional printing (Liu et al., 2021b).

REFERENCES

Beatrice, C., F. Fiorillo, F. J. Landgraf, V. Lazaro-Colan, S. Janasi, and J. Leicht. 2008. Magnetic loss, permeability dispersion, and role of eddy currents in Mn-Zn sintered ferrites. *Journal of Magnetism and Magnetic Materials* 320(20): e865–8.

Birčáková, Z., J. Füzer, P. Kollár, J. Szabó, M. Jakubčin, M. Streckova, R. Bureš, and M. Fáberová. 2020. Preparation and characterization of iron-based soft magnetic composites with resin bonded nano-ferrite insulation. *Journal of Alloys and Compounds* 828:154416.

Burdett, J. K., G. D. Price, and S. L. Price. 1982. Role of the crystal-field theory in determining the structures of spinels. *Journal of the American Chemical Society* 104(1):92–5.

Chen, Z., X. Sun, Z. L. Ding, and Y. Q. Ma. 2018, Manganese ferrite nanoparticles with different concentrations: preparation and magnetism. *Journal of Materials Science & Technology* 34(5):842–7

Coey, J. M. 2010. *Magnetism and magnetic materials*. New York: Cambridge University Press.

Costa, A. C. F. M., E. Tortella, M. R. Morelli, and R. H. G. A. Kiminami. 2003. Synthesis, microstructure and magnetic properties of Ni-Zn ferrites. *Journal of Magnetism and Magnetic Materials* 256(1–3):174–82.

Ehrhardt, H., S. J. Campbell, and M. Hofmann. 2003. Magnetism of the nanostructured spinel zinc ferrite. *Scripta Materialia* 48(8):1141–6.

Fiorillo, F., C. Beatrice, O. Bottauscio, and E. Carmi. 2014. Eddy-current losses in Mn-Zn ferrites. *IEEE Transactions on Magnetics* 50(1):6300109.

Gore, S. K., R. S. Mane, M. Naushad, S. S. Jadhav, M. K. Zate, Z. A. Alothman, and B. K. N. Hui. 2015. Influence of Bi^{3+}-doping on the magnetic and Mossbauer properties of spinel cobalt ferrite. *Dalton Transactions* 44(14):6384–90.

Ishino, K., and Y. Narumiya. 1987. Development of magnetic ferrites-control and application of losses. *American Ceramic Society Bulletin* 66(10):1469–74.

Jalaly, M., M. H. Enayati, and F. Karimzadeh. 2009. Investigation of structural and magnetic properties of nanocrystalline $Ni_{0.3}Zn_{0.7}Fe_2O_4$ prepared by high energy ball milling. *Journal of Alloys and Compounds* 480(2):737–40.

Jenus, P., A. Ucakar, S. Repse, C. Sangregorio, M. Petrecca, M. Albino, R. Cabassi, C. de Julian Fernandez, and B. Belec. 2021. Magnetic performance of

$SrFe_{12}O_{19}$-$Zn_{0.2}Fe_{2.8}O_4$ hybrid magnets prepared by spark plasma sintering. *Journal of Physics D: Applied Physics* 54(20):204002.

Kharisov, B. I., H. V. Rasika Dias, and O. V. Kharissova. 2019. Mini-review: ferrite nanoparticles in the catalysis. *Arabian Journal of Chemistry* 12 (7):1234–1246.

Kotnala, R. K., and J. Shah. 2015. Ferrite materials: nano to spintronics regime. In *Handbook of magnetic materials*, ed. E. Brück, 23:291–379. Amsterdam: North-Holland Publishing.

Liu, B. B., L. Zhang, B. Zhang, J. Wang, Y. B. Zhang, G. H. Han, and Y. J. Cao, 2021a. Characterizations on phase reconstruction, microstructure evolution and separation of magnetic ferrite ceramics from low-grade manganese ores by novel uphill reaction diffusion and magnetic separation. *Materials Characterization* 175:111028.

Liu, L. B., K. D. T. Ngo, and G. Q. Lu. 2021b. Effects of SiO_2 inclusions on sintering and permeability of NiCuZn ferrite for additive manufacturing of power magnets. *Journal of the European Ceramic Society*. 41(1):466–71.

Manju, B. G., and P. Raji. 2018. Synthesis and magnetic properties of nano-sized $Cu_{0.5}Ni_{0.5}Fe_2O_4$ via citrate and aloe vera: a comparative study. *Ceramics International* 44(7):7329–33.

Mugutkar, A. B., S. K. Gore, U. B. Tumberphale, V. V. Jadhav, R. S. Mane, S. M. Patange, S. F. Shaikh, M. Ubaidullah, A. M. Al-Enizi, and S. S. Jadhav. 2020. The role of La^{3+} substitution in modification of the magnetic and dielectric properties of the nanocrystalline Co-Zn ferrites. *Journal of Magnetism and Magnetic Materials* 502:166490.

Nakamura T., T. Miyamoto, and Y. Yamada. 2003. Complex permeability spectra of polycrystalline Li-Zn ferrite and application to EM-wave absorber. *Journal of Magnetism and Magnetic Materials* 256(1–3):340–7.

Otobe, S., Y. Yachi, T. Hashimoto, T. Tanimori, T. Shigenaga, H. Takei, and K. Hontani. 1999. Development of low loss Mn-Zn ferrites having the fine microstructure. *IEEE Transactions on Magnetics* 35(5):3409–11.

Ott, G., J. Wrba, and R. Lucke. 2003. Recent developments of Mn-Zn ferrites for high permeability applications. *Journal of Magnetism and Magnetic Materials* 254:535–7.

Pardavi-Horvath, M. 2000. Microwave applications of soft ferrites. *Journal of Magnetism and Magnetic Materials* 215:171–83.

Phumying, S., S. Labuayai, E. Swatsitang, V. Amornkitbamrung, and S. Maensiri. 2013. Nanocrystalline spinel ferrite (MFe_2O_4, M = Ni, Co, Mn, Mg, Zn) powders prepared by a simple aloe vera plant-extracted solution hydrothermal route. *Materials Research Bulletin* 48(6):2060–5.

Pullar, R. C. 2012. Hexagonal ferrites: a review of the synthesis, properties and applications of hexaferrite ceramics. *Progress in Materials Science* 57(7):1191–334.

Shirsath, S. E., D. Y. Wang, S. S. Jadhav, M. L. Mane, and S. Li. 2018. Ferrites obtained by sol-gel method. In *Handbook of sol-gel science and technology*, ed. L. Klein, M. Aparicio, and A. Jitianu, 695–735. Switzerland: Springer International Publishing.

Shokrollahi, H. 2008. Magnetic properties and densification of manganese-Zinc soft ferrites ($Mn_{1-x}Zn_xFe_2O_4$) doped with low melting point oxides. *Journal of Magnetism and Magnetic Materials* 320(3–4):463–74.

Shokrollahi, H., and K. Janghorban. 2007. Soft magnetic composite materials (SMCs). *Journal of Materials Processing Technology* 189(1–3):1–12.

Smit, J., and H. P. J. Wijn. 1959. *Ferrites: physical properties of ferrimagnetic oxides in relation to their technical applications*. Hoboken, NJ: John Wiley Sons.

Snoek, J. L. 1949. *New developments in ferromagnetic materials: with introductory chapters on the statics and the dynamics of ferromagnetism*. New York: Elsevier Publishing.

Soliman, S., A. Elfalaky, G. H. Fecher, and C. Felser. 2011. Electronic structure calculations for $ZnFe_2O_4$. *Physical Review B* 83(8):085205.

Sugimoto, M. 1999. The past, present, and future of ferrites. *Journal of the American Ceramic Society* 82(2):269–80.

Sun, K., Y. P. Li, S. Feng, Q. Z. Gao, Z. X. Wang, X. M. Wei, L. C. Ju, and R. H. Fan. 2021. Optimizing the soft magnetic properties of Mn-Zn ferrite by a proper control of sintering process. *Journal of Electronic Materials* 50 (3):1467–73.

Szotek, Z., W. M. Temmerman, D. Koedderitzsch, A. Svane, L. Petit, and H. Winter. 2006. Electronic structures of normal and inverse spinel ferrites from first principles. *Physical Review B* 74(17):174431.

Thakur, P., C. Deepika, T. Shilpa, B. Nikhil, and T. Atul. 2020. A review on MnZn ferrites: synthesis, characterization and applications. *Ceramics International* 46(10):15740–63.

van der Zaag, P. J. 1999. New views on the dissipation in soft magnetic ferrites. *Journal of Magnetism and Magnetic Materials* 196:315–9.

van der Zaag, P. J. 2021. Ferrites. In *Encyclopedia of materials: science and technology* (Second Edition), ed. K. H. Jürgen Buschow, R. W. Cahn, M. C. Flemings, B. Ilschner, E. J. Kramer, S. Mahajan, and P. Veyssière, 3033–7. Oxford: Pergamon Press.

Vangroenou, A. B., P. F. Bongers, and A. L. Stuyts. 1969. Magnetism microstructure and crystal chemistry of spinel ferrites. *Materials Science and Engineering* 3(6):317–92.

Vedrtnam, A., K. Kishor, D. Sunil, and K. Aman. 2020. A comprehensive study on structure, properties, synthesis and characterization of ferrites. *Aims Materials Science* 7(6):800–35.

Yan, M., S. B. Yi, X. Y. Fan, Z. H. Zhang, J. Y. Jin, and G. H. Bai. 2021. High-frequency MnZn soft magnetic ferrite by engineering grain boundaries with multiple-ion doping. *Journal of Materials Science & Technology* 79:165–70.

Yang, B., and Z. Wang. 2016. The structure and magnetic properties of NiCuZn ferrites sintered via a two-step sintering process. *Journal of Sol-Gel Science and Technology* 80(3):840–7.

FURTHER READING

Goldman, A. 2006. *Modern ferrite technology* (Second Edition). Boston, MA: Springer Science and Business Media, Inc.

IV

Other Functional Magnetic Materials

CHAPTER 13

Materials with Magnetic-X Effects

13.1 MAGNETO-OPTICAL MATERIALS

13.1.1 Magneto-Optical Effects

Since the discovery of Faraday effect by the British physical scientist Michael Faraday in 1845 (Faraday, 1846), this wonderful interaction between light and magnetism has ignited extensive research. When an external magnetic field is applied, the electromagnetic properties of magnetic material change, so as the transmission characteristics of light waves within it. Such phenomena are called the magneto-optical effects, among which the Faraday effect and Kerr effect are the most well-known and widely used magneto-optical effects (McCord, 2015).

For the Faraday effect, the polarized plane of a light beam rotates by an angle θ_F during transmission through an at least partly transparent magnetized material with the magnetic field H parallel to the light propagation direction, as shown in Figure 13.1. The Faraday rotation angle θ_F can be described as

$$\theta_F = VHL \tag{13.1}$$

where V denotes the Verdet constant of a Faraday material, and L is the transmission path length. Faraday effect is unique due to its non-reciprocal characteristic, meaning that the reversed propagation direction of the light wave across the same magnetic field region induces the analogous rotation along the same direction in Figure 13.1.

Unlike the Faraday effect in the transmission geometry, the Kerr effect belongs to the reflection geometry, meaning that the polarized plane of an incident light rotates by a certain angle θ_k during reflection from a magnetized and reflective substance, as schematically shown in Figure 13.2a. According to the magnetization direction relative to the incidence and reflection plane of the light beam, Kerr effect can be divided into three basic configurations: polar, longitudinal and transverse. For the polar Kerr effect, magnetization M of the medium sample is perpendicular to the reflecting surface, as shown in Figure 13.2b. For the longitudinal and transverse Kerr effects, M lies in the surface plane parallel or perpendicular to the incident plane (Figures 13.2c and d). In general, the polar Kerr effect gives rise to larger polarization changes compared with the longitudinal one, while there is no obvious magnetic rotation in the transverse direction (Qiu and Bader, 2000). As introduced in Chapter 3, the magneto-optical effects have been widely utilized for magnetic measurements (Higashida et al., 2020; Aoshima et al.,

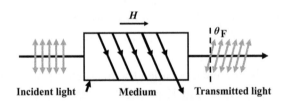

FIGURE 13.1 Schematic diagram of the Faraday effect.

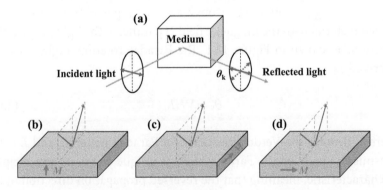

FIGURE 13.2 (a) Schematic diagram of the Kerr effect, (b) polar, (c) longitudinal and (d) transverse Kerr effect.

2018). For example, the in-situ magnetic domain observation using the Kerr microscopy is one of the most important applications of the Kerr effect (Zhang et al., 2018b).

13.1.2 Materials Based on Magneto-Optical Effects

Magneto-optical materials that function over ultraviolet to infrared band can be classified into glass, crystal and ceramic ones. The magneto-optical glass with superior size scalability can be used in large diameter glass laser systems, but its application in high-power lasers is restricted by the low thermal conductivity. Compared to the glass, the magneto-optical crystal possesses superior properties, including high thermal conductivity, large laser-induced-damage threshold and large Verdet constant, but with size limitations. While the magneto-optical ceramics exhibit the advantages of simple preparation process and low cost over the other alternatives, their magneto-optical performance is relatively inferior. In the following, typical magneto-optical glass, crystal and ceramic materials will be described.

13.1.2.1 Magneto-Optical Glass

Depending on the dopants with different electron configurations, the magneto-optical glass exhibits paramagnetic or diamagnetic responses. The paramagnetic glass containing rare-earth (RE) ions such as Tb^{3+} (Yin et al., 2018), Eu^{2+} (Tanaka et al., 1997), Nd^{3+} (Golis, 2016), Pr^{3+} and Dy^{3+} (Malakhovskii et al., 2003). The Verdet constant as a quantitative measure of the magneto-optical capability is weak in most materials, but increases significantly for the REs-doped magneto-optical glass. The paramagnetism arises from the unpaired electrons in the REs orbitals as represented by the non-zero angular momentum J. The large Verdet constant enables highly sensitive characteristic for the paramagnetic glass, but is inversely proportional to the temperature (Potseluyko et al., 2003).

The diamagnetic glass incorporated with heavy-metal species such as Bi^{3+}, Pb^{2+}, Te^{4+} and Sb^{3+} exhibits large polarizability (Chen and Ma, 2020; Elisa et al., 2017). Although the Verdet constant is temperature independent, the stable diamagnetic glass with small Verdet constant and low magnitude of Faraday effect requires increased optical path length L for sufficient Faraday rotation angle θ_F, based on Formula (13.1). As such the extra material cost and linear birefringence in practical applications become unavoidable shortcomings for the diamagnetic glass, compared to their paramagnetic alternatives (Shen et al., 2018).

Among different dopant species, the paramagnetic Tb^{3+} with large magnetic moments among all lanthanides (angular momentum $J = 6$, Landé splitting factor $g = 1.46$) has received the greatest attention. As of now, all commercially available magneto-optical glasses rely on the massive doping of Tb^{3+} (Schmidt et al., 2011). An efficient paramagnetic system is 50 mol% Tb_2O_3/Dy_2O_3 co-doped B_2O_3-Ga_2O_3-SiO_2-P_2O_5 borate glass with high Verdet constant of −185.3 rad/T/m at the wavelength of 632.8 nm and external magnetic field of 1 T (Hayakawa et al., 2002). Upon a higher content of Tb_2O_3 to 60 mol%, Suzuki et al. have successfully achieved a larger Verdet constant of −234 rad/T/m for borate glass (Suzuki et al., 2018).

To sum up, the magneto-optical glass has broad application prospects in magneto-optical isolators, modulators, diodes and optical current sensors. Despite the benchmark crystal materials, the magneto-optical glass with high RE and heavy-metal solubility exhibits significant potential for applications in fiber-integrated devices.

13.1.2.2 Magneto-Optical Crystals

Deeper understandings on the crystal growth theory, together with the continual advancement of processing technology have yielded breakthroughs in the field of magneto-optical crystals during the past decades. The majority of current magneto-optical applications are based on the garnet-type rare-earth ferrite crystals to provide high Verdet constants and unique magneto-optical performance. The benchmark materials include the yttrium iron garnet (YIG, $Y_3Fe_5O_{12}$), terbium aluminum garnet (TAG, $Tb_3Al_5O_{12}$) and terbium gallium garnet (TGG, $Tb_3Ga_5O_{12}$) (Geho et al., 2004).

Since first discovery in 1958 (Dillon, 1958), the garnet-type YIG with large Faraday rotation has been used in infrared or near-infrared range such as for optical telecommunications. It belongs to the cubic crystal structure, containing 8 $Y_3Fe_5O_{12}$ in each unit cell. From the general formula $Y_3Fe_2Fe_3O_{12}$, there are two non-equivalent Fe^{3+} sites, where the 16a and 24b sites are octahedral or tetrahedral coordinated with the oxygen anions (Akhtar et al., 2016). The trivalent Y^{3+} ions occupy the 24c sites and can be substituted with other REs such as Ce, Nd and Gd.

As the YIG crystal is an incongruent melting compound, the Czochralski (CZ) method is not applicable for the crystal growth. The traditional flux method is not only time consuming, but also difficult to control the grain growth. Although new methods such as the modified floating-zone method and the laser-heated floating-zone method have been introduced, it remains difficult to prepare large-size crystals. The edge-defined film-fed growth method has been reported to enable faster growth rate of the Ga:YIG crystals at effectively reduced cost (Zhuang et al., 2013). Techniques such as metal-organic chemical vapor deposition, liquid phase epitaxy and reactive radiofrequency magnetron sputtering have been utilized to prepare the Bi-doped (Ishibashi et al., 2005) and Ce-doped (Srinivasan et al., 2020) rare-earth garnet thin films with merits of enhanced magneto-optical effect and high magnetic sensitivity, which are compatible for integrated silicon photonics in microscale chips.

For the shorter wavelength band below 1100 nm (Chen et al., 2015b), the YIG with rather poor light transmittance and intense optical absorption are not applicable. Hence new magneto-optical crystals TAG and TGG have been explored. Although the TAG crystal exhibits higher Verdet constant than the mainstream TGG, its incongruent melting nature makes it difficult to be grown by the CZ method (Geho et al., 2005). The floating method can only be utilized to produce small TAG single crystals, limiting the size for practical applications (Man et al., 2017). Comparably, the TGG crystal with large size and high optical quality via the CZ method dominates the market. Current TGG crystals with a diameter of ~50 mm are relatively common, which are much larger than other counterparts. It has been reported that the optical transmittance of the TGG crystal ($\Phi 40 \times 5$ mm^3) can be beyond 90% in the visible-near-infrared range, and the Verdet constant reaches 196.5 rad/T/m at the wavelength of 532 nm (Jin et al., 2018).

13.1.2.3 Magneto-Optical Ceramics

As mentioned above, the growth of high-performance TAG single crystal from the melt remains an extremely difficult task. In recent years, the mature preparation methods of transparent ceramics, however, provide possibility to synthesize the TAG magneto-optical ceramics. During the solid phase reaction sintering process, REs dopants including Y^{3+} (Chen et al., 2012), Ce^{3+} (Chen et al., 2015a; Starobor et al., 2014), Tm^{3+} (Hao et al., 2019) ions, as well as sintering aids including ZrO_2 (Chen et al., 2019), SiO_2 (Hao et al., 2018) and MgO (Zhang et al., 2018a) have been proved to improve the magneto-optical properties of TAG ceramics. For example, large-scale and full-dense $(Tb_xY_{1-x})_3Al_5O_{12}$ magneto-optical ceramics ($\Phi 10 \times L40$ mm^3 or $\Phi 45 \times t10$ mm^3) have been prepared by solid-state reaction, followed by hot isostatic pressing treatment, which are suitable for high-power laser systems over kW (Aung and Ikesue, 2017). The optical scattering loss is about 10^2–10^4 times superior to that of the TAG crystal grown by the floating-zone approach (Aung and Ikesue, 2017). Simultaneously, the measured Verdet constant of 307 rad/T/m at the wavelength of 532 nm is approximately 1.5 times of the value for commercial TGG crystal grown by the CZ approach (Aung and Ikesue, 2017).

13.1.3 Applications

Since the discovery of Faraday magneto-optical effect in 1845 (Faraday, 1846), the exploration for the magneto-optic interaction has continued for centuries. However, it is until the last three decades that the application of magneto-optical effects emerges, ranging from optoelectronics, optical communication, to other frontier sciences. The following part provides a glimpse of the common magneto-optical applications, including magneto-optical recording, modulator, isolator and switcher.

13.1.3.1 Magneto-Optical Recording

The magneto-optical recording integrates the optical recording and magnetic recording, with the high storage density, removability and nonvolatility functions that have developed as an industrially important information storage technique (Tsunashima, 2001). To qualify as a magneto-optical recording medium for massive application, the following criteria should be satisfied. Firstly, the magnetic anisotropy should be perpendicular to the film plane, exhibiting a rectangular hysteresis loop and high coercivity that drops sharply when illuminated with a focused laser beam. Secondly, high magneto-optical recording sensitivity (reduced laser recording power), large magneto-optical effect (large Faraday or Kerr angle) and low disk writing noise are required. Thirdly, satisfactory thermal stability, chemical stability and low-cost production are also necessary. There are three main types of magneto-optical recording materials including manganese-bismuth (MnBi) film (Furuya et al., 2016), amorphous rare-earth-transition metal (RE-TM) film (Murakami and Birukawa, 2003), and rare-earth iron garnet (Lomako, 2013; Nakamura et al., 2014).

13.1.3.2 Magneto-Optical Modulator

For the magneto-optical modulator, the polarization plane of light rotates through the magneto-optical materials under the excitation of magnetic field to modulate the signal light beam. Generally, the magneto-optical modulator consists of polarizers, magnetic field generator, magneto-optical material, together with the analyzer. The magneto-optical modulators have found wide applications in modulating TV signals, as well as diverse optical detection systems for both civil and military applications. The Ce/Ga co-doped $Gd_3Fe_5O_{12}$ single crystal thin film via improved edge-defined film-fed growth method exhibits a Faraday rotation angle as high as ~10^3 Deg/cm with improved optical transmittance (Liu et al., 2021). Through special intracavity power-modulator of a $Nd:YVO_4$ using the TGG magneto-optical effect, the transmittance of the cavity changes periodically by changing the magnetic intensity, and the output power can also be modulated (Liu et al., 2020).

13.1.3.3 Magneto-Optical Isolator

An optical isolator, also known as unidirectional device, is a non-reciprocal device that permits light to pass along only one direction. Such device can be used to eliminate reflected light generated in optical fiber transmission, and effectively improve the stability of laser diodes emission and optical circulators. It has become an essential component in high-quality and high-power optical systems (Abadian et al., 2021). Magneto-optical isolators can be made of magneto-optical single crystals or magneto-optical films. Among the magneto-optical materials, YIG with large Faraday rotation and low loss is a preferable choice for industrialization (Stadler and Mizumoto, 2014).

13.1.3.4 Magneto-Optical Switcher

The magneto-optical switcher uses the Faraday effect to control the rotation of the polarization plane of the transmitted light by tuning the magnetization of magneto-optical crystal via the external magnetic field, so as to switch the optical pathway (Ishida et al., 2016; Haddadpour et al., 2016). The magneto-optical switchers can be made in the forms of blocks, films and fibers. Compared with electro-optical or thermo-optical switchers, magneto-optical switchers exhibit numerous advantages such as fast speed, low hysteresis, low energy consumption, satisfactory stability, low driving voltage, small size and easy integration. The disadvantage of magneto-optical switchers, however, lies in the fact that the magneto-optical performance needs to be further improved for increased light transmittance and reduced loss. Recently, based on the Ce-substituted YIG film, a novel self-holding magneto-optical waveguide switch controlled by a 1-µs pulsed electric current has been designed, exhibiting an extinction ratio up to 15.4 dB (Okazeri et al., 2018).

13.2 MAGNETOSTRICTIVE MATERIALS

13.2.1 Magnetostriction

The magnetostrictive effect manifests itself by the deformation of ferromagnetic or ferrimagnetic materials under varied magnetization states (Lee, 1955). As illustrated in Figure 13.3, when approaching the Curie temperature T_C, materials with magnetic transition between ferromagnetic and paramagnetic states undergo spontaneous deformation, known as the spontaneous magnetostriction. Below the T_C, when imposing a magnetic field, the initially disordered magnetic moments uniformly orient toward the direction of the external field H, causing changes in the volume and length, which are called volume and linear magnetostriction, respectively. Compared with the linear magnetostriction, the volume magnetostriction is much weaker. To characterize the magnetostrictive effect of a given material, magnetostriction coefficient λ is used, which can be written as

$$\lambda = \Delta l / l \qquad (13.2)$$

where Δl is length change along the same direction of the initial length l, as shown in Figure 13.4. Here $\lambda > 0$ represents that the material elongates along the direction of H, i.e. positive magnetostriction as exhibited in Fe.

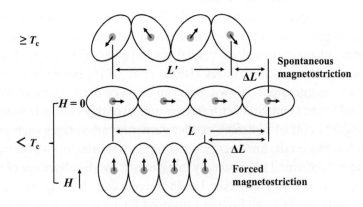

FIGURE 13.3 Schematic diagram of the magnetostriction mechanism.

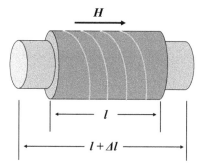

FIGURE 13.4 Schematic diagram of the longitudinal magnetostriction.

Accordingly, $\lambda < 0$ means that the material contracts along the direction of H, i.e. negative magnetostriction as exhibited in Ni. Zero thermal expansion and dimensional stability can be achieved for the well-known $Fe_{64}Ni_{36}$ Invar alloy, catering to precision apparatuses and instruments.

13.2.2 Materials Based on Magnetostrictive Effects

The magnetostrictive effect was first discovered by Joule in 1842, finding that the length of iron bar changed when applied an external magnetic field. In the 1860s, Villari discovered stress-induced magnetization variation, i.e. the applied mechanical load induces the magnetic domain rotation and changes the magnetization. Such inverse effect of magnetostriction is known as the Villari effect. Note that traditional magnetostrictive material with small λ in the order of several ppm (similar to the dimension of uncontrolled thermal deformation) has few engineering applications. Such situation has been changed much later when the giant magnetostrictive material (GMM) with saturation magnetostriction λ_s beyond 30 ppm was developed. In the 1960s, RE metals including Tb and Dy have been identified with significantly larger magnetostriction (~10000 ppm) than that of Ni (Legvold et al., 1963; Clark et al., 1965), which unfortunately are not feasible for practical application due to the ultra-low T_C below room temperature. Continual attempts lead to the discovery of $REFe_2$ compound (RE = Sm, Tb and Dy) with simultaneous high T_C and giant room-temperature λ_s (Clark and Belson, 1972). To minimize the magnetocrystalline anisotropy with maintained high magnetostriction (>2000 ppm), Clark et al. have developed the $Tb_{0.3}Dy_{0.7}Fe_2$ compound (commercially known as the Terfenol-D) in 1972 (Clark et al., 1972). However, two main drawbacks of

TABLE 13.1 Specifications of Terfenol-D and Galfenol (Xue et al., 2018)

Parameters	Terfenol-D	Galfenol
Chemical composition	$Tb_xDy_{1-x}Fe_2$ ($x = \sim 0.3$)	$Fe_{1-x}Ga_x$ ($x = 0.12$–0.33)
Conductibility (S/m)	1.72×10^6	5.96×10^6
Relative permeability	3–10	30–700
Young modulus (GPa)	35–65	60–80
Bulk density (kg/m^3)	9250	7870
Tensile strength (MPa)	28	500
Static hysteresis loss (kJ/m^3)	23	3.1
Saturated magnetostriction	2000×10^{-6}	395×10^{-6}
Curie temperature (°C)	380	700
Saturation required field (kA/m)	≥ 65	≤ 24

Terfenol-D retard its large-scale production, including its brittle nature and heavy reliance on the critical REs Dy/Tb. Later on, $Fe_{1-x}Ga_x$ alloy ($x = 0.12$–0.33, generally termed as Galfenol) with moderate magnetostriction ($\sim$395 ppm at room temperature) has been developed (Arthur and Clark, 2002). From the specifications in Table 13.1, the Galfenol possesses outstanding tensile strength ($\sim$500 MPa) and machinability, together with high T_C and low cost. In the following, Terfenol-D and Galfenol as the two most widely used GMMs will be further discussed.

13.2.2.1 Terfenol-D

Terfenol-D is a widely used magnetostrictive material with the composition of $Tb_{0.3}Dy_{0.7}Fe_2$. For the AB_2-type cubic Laves phase, the unit cell consists of 8 RE atoms and 16 TM atoms (Liu et al., 2012). In 1977, a magnetostrictive atomic model based on the Laves phase lattice structure has been proposed, explaining the higher λ_{111} than λ_{100} (Busbridge and Piercy, 1995; Wun-Fogle et al., 1998). At room temperature, the magnetostriction λ_{111} origins from the rhombohedral distortion of the cubic lattice ($\sim$2400 ppm). When approaching the spin reorientation temperature ($\sim$285 K), the easy axis changes from [111] to [100], accompanied with the crystallographic transformation from rhombohedral to tetragonal (Wang et al., 2016). The magnetostriction λ_{100} from the tetragonal distortion is limited at $\sim$100 ppm. Consequently, many efforts have been devoted to fabricating perfect Terfenol-D <111> single crystals to maximize the magnetostriction, among which directional solidification technology (Bridgman method, floating-zone method and Czochralski method) has been widely investigated. Heat treatment has also been employed to improve the comprehensive performance via optimizing the distribution of RE-rich phase. Besides, technical advances in alloying strategy are progressing, including Tb or Dy replacement by cheaper Nd, Pr or Ho, and Fe replacement by Co, Al, Mn, Si or Zr. For $(Tb_{0.3}Dy_{0.7})_{1-x}Pr_xFe_{1.55}$ alloys ($0 \leq x \leq 0.35$), the lattice parameter a of the Laves phase increases while T_C decreases with raised Pr content, and the linear anisotropic magnetostriction λ_a ($\lambda_a = \lambda_\| - \lambda_\perp$) exhibits a maximum of 1502 ppm at $x = 0.25$ (Lv et al., 2009). Minor Co addition has been found to stabilize the Laves phase and increase the T_C, at the sacrifice of magnetostriction coefficient (Bodnar et al., 2010). Other elements such as Al (Xie et al., 2000), Si (Xu et al., 2008) and Cr (Nolting and Summers, 2015) have also been studied to increase the magnetostrictive properties of Terfenol-D alloy.

13.2.2.2 Galfenol

To mitigate the worldwide RE criticality, particularly for the scarce Tb/Dy in Terfenol-D material, Galfenol as a non-RE alternative has been developed with prospective applications. Although its λ_s is inferior to the Terfenol-D, an order of magnitude lower field is required to saturate the magnetostriction. For the FeGa alloy, incorporating 15.0–27.5 at% Ga into the Fe lattice improves the magnetostriction by tenfold. The solubility of Ga in bcc α-Fe crystal lattice highly affects the magnetostrictive performance (Clark et al., 2003; Muñoz-Noval et al., 2018), as evidenced by the two peaks of λ near 19 at% and 27 at% Ga for the quenched FeGa alloy (Figure 13.5).

The origin of high λ in Galfenol has been a research hotspot. From the equilibrium and metastable phase diagram of FeGa, multiple phase structures can coexist under different thermal treatments, including the solid-solution disordered A2, as well as the ordered B2, D0$_3$, D0$_{19}$ and L1$_2$, which increase the complexity in correlating the microstructure with the magnetostrictive behaviors (Srisukhumbowornchai and Guruswamy, 2002). Numerous theoretical and experimental investigations have been conducted to unveil the structure-property dependence of FeGa alloys, giving rise to two models. For the first model, the peak near 19 at% Ga has been attributed to short-range Ga-Ga atom pairs in the [100] axis of the bcc A2 structure, and increased magnetoelastic coupling, while the other peak near 27 at% Ga is due to softening of the shear modulus (Wu et al. 2003; Clark et al., 2001; 2003). For the second extrinsic model proposed by

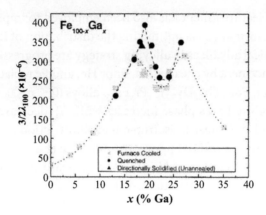

FIGURE 13.5 $(3/2)\lambda_{100}$ as a function of Ga concentration for the $Fe_{100-x}Ga_x$ alloys. Reprinted with permission from (Clark et al., 2003). Copyright (2003) AIP Publishing.

Khachaturyan and Viehland, it is based on the heterogeneous microstructure, i.e. nanoscale clusters within the bcc phase lattice significantly contribute to the macroscopic magnetostriction, which also accounts for the reported thermal history dependence of magnetostriction (Khachaturyan and Viehland, 2007; Boisse et al., 2011).

The large magnetostriction of FeGa alloy is dependent on the [100] texture, which is also the easy growth direction and easy magnetization direction. Consequently, the main preparation method of FeGa alloy is directional solidification, involving the Bridgman method, the Float zone melting method and the Czochratski method. A directionally growth rate of 22.5 mm/h generates the polycrystalline $Fe_{72.5}Ga_{27.5}$ alloy with a near [100] texture and a large λ_s of 271 ppm (Srisukhumbowornchai and Guruswamy, 2001). Under an optimum directional solidification condition, satisfactory tensile fracture strain of 12.5% and λ_s of 387 ppm have been achieved for the $Fe_{81}Ga_{19}$ alloy doped with trace amount of Tb (0.05 at%) (Wu et al., 2019). Incorporating the third component Pr (0.20 at%) has also been reported to increase the magnetocrystalline anisotropy and strengthen the transverse magnetostriction up to 800 ppm (He et al., 2016).

13.2.3 Applications

Since the discovery of room-temperature magnetostrictive effect, the magnetostrictive materials have developed rapidly for various applications (Apicella et al., 2019). The Joule or direct magnetostriction with observable deformation as a response to the magnetic field H can be exploited for

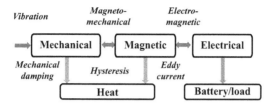

FIGURE 13.6 Energy flow within a magnetostrictive energy harvester. Reprinted with permission from (Deng and Dapino, 2017). Copyright (2017) IOP Publishing.

actuation (Deng and Dapino, 2017). The inverse Villari effect as a response to the applied mechanical stress and resultant magnetization M caters to sensing/harvesting (Narita and Fox, 2018). The Wiedemann effect refers to the twisting of a magnetostrictive cylinder caused by simultaneous longitudinal and circumferential magnetic fields H_l and H_c, which can be used for actuation/torque (Vinogradov et al., 2017). The Matteucci effect, an inverse effect of Wiedemann effect, can be applied for sensing.

Magnetostrictive actuators with high energy density and intrinsic robustness boost the applications in smart materials and devices, such as micropositioning, micro- or inchworm-motors, active vibration control and micro-pumps. According to Table 13.1, Terfenol-D with high λ and high output may deliver sufficiently high displacement and force simultaneously, while Galfenol with relatively lower λ is more suitable for micro-electromechanical systems. Magnetostrictive materials are also widely used in vibration energy harvesters, with the related mechanism displayed in Figure 13.6. The magneto-mechanical coupling of magnetostrictive material transforms the mechanical energy into the magnetic energy, which is then converted into the electrical energy through the electromagnetic coupling (Deng and Dapino, 2017). Magnetostrictive materials may also be used for interdisciplinary fields such as bio-sensing for the detection of bacterial spores, proteins, classical swine fever and COVID-19 (Narita et al., 2021). Continuous and in-depth research is encouraging to apply novel magnetostrictive materials for more emerging nanotechnologies.

13.3 MAGNETOCALORIC MATERIALS

13.3.1 Magnetocaloric Effect

Magnetocaloric effect refers to the adiabatic temperature change of a magnetic material exposed to alternating magnetic field, i.e. the temperature rises upon applied magnetic field (heating process) and drops when the field is removed (cooling process). Figure 13.7 displays the schematic magnetocaloric effect during the adiabatic magnetization and demagnetization cycle (Chaudhary et al., 2019). As shown in Figures 13.7a and b, the application of magnetic field H transforms the disordered magnetic moment to an ordered state, decreasing the magnetic entropy ($\Delta S < 0$). Under the isentropic condition with constant total entropy (adiabatic process), the decreased magnetic entropy is compensated by the lattice entropy, as characterized by the increased temperature ($T + \Delta T$, $\Delta T > 0$). With the heat-transfer fluid, the temperature of the system reduces back to the initial temperature T (Figure 13.7c). The removal of magnetic field H increases the magnetic entropy ($\Delta S > 0$), hence decreasing the lattice entropy and the temperature (Figure 13.7d). The above expelled and absorbed heat cycle during cyclic application and removal of magnetic field may construct a basic magnetic refrigerator, including the main components of magnetocaloric material, magnetic field generator, heat exchangers and heat transfer systems.

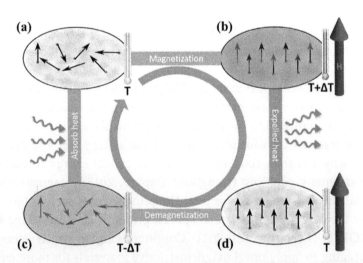

FIGURE 13.7 Schematic magnetocaloric cycle of the magnetocaloric material. (a) Initial unmagnetized state with random magnetic moments; (b) adiabatically magnetized state with increased temperature; (c) back to the initial temperature with expelled heat; and (d) adiabatically demagnetized state with decreased temperature. Reprinted with permission from (Chaudhary et al., 2019). Copyright (2019) Elsevier.

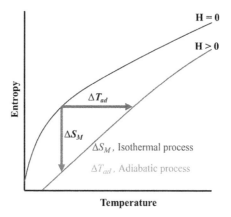

FIGURE 13.8 A schematic of the entropy of magnetocaloric material as function of temperature. Reprinted with permission from (Chaudhary et al., 2019). Copyright (2019) Elsevier.

The magnetocaloric effect is intrinsic, which can be expressed in two ways, one with the adiabatic temperature change ΔT_{ad} (shown by the horizontal arrow in Figure 13.8), the other with the isothermal entropy change ΔS_M (shown by the vertical arrow in Figure 13.8). Both processes can be defined as (Law, 2012)

$$\Delta T_{ad} = \int_0^H -\frac{T}{C_H}\left(\frac{\partial M}{\partial T}\right)_H dH \qquad (13.3)$$

$$\Delta S_M = \int_0^H \left(\frac{\partial M}{\partial T}\right)_H dH \qquad (13.4)$$

where C_H represents the specific heat capacity. The ΔT_{ad} and ΔS_M are two main figures of merit to evaluate the magnetocaloric effect. Between the two parameters, the metric ΔT_{ad} is more intuitive and straightforward, while the ΔS_M is easier to measure experimentally. Based on Equation (13.3), it is evident that the giant magnetocaloric effect is dependent on the large changes of magnetization with temperature, as well as the low C_H of the magnetocaloric materials. Considering the abrupt drop of magnetization for ferromagnetic materials when approaching the T_C, the absolute value of ΔS_M can be maximized with $T = T_C$. Consequently, the magnetocaloric material exhibiting large magnetocaloric effect with T_C close to room temperature has attracted intensive research interests.

13.3.2 Materials Based on Magnetocaloric Effect

Compared to the traditional vapor-compression refrigeration technology with harmful chlorofluorocarbon and carbon dioxide emissions, the magnetocaloric materials provide an alternative solution with environmentally friendly and highly efficient characteristics. The history of magnetocaloric materials can be traced back to 1917 with the finding of heated Ni under changing magnetic field by Weiss and Piccard (Weiss and Piccard, 1917). Later on, Debye and Giauque theoretically interpreted the thermodynamic magnetocaloric effect, which was then confirmed by the sub-Kelvin cooling through adiabatic demagnetization of paramagnetic salt (Debye, 1926; Giauque, 1927; Giauque and MacDougall, 1933). However, it was until the late 1970s that the first prototype of magnetic refrigeration was achieved by the operation of metallic Gd near room temperature, which has also initiated the search for novel magnetocaloric materials (Kitanovski, 2020). To date, several classes of room-temperature or near room-temperature magnetocaloric materials have been found, including the Gd-based alloys, Mn-based alloys, Ni-Mn-based Heusler alloys and LaFeSi alloys.

13.3.2.1 Gd-Based Alloys

In 1976, the RE metal Gd has been reported to exhibit large temperature increment of 14 K under adiabatic magnetization condition (7 T) near the Curie temperature T_C (Brown, 1976). Among different RE metals, Gd is a RE element with T_C of 293 K (approaching the room-temperature region) and a large magnetocaloric effect of 9.5 J/kg·K, which opens the door to investigate the room-temperature magnetic refrigeration. Figure 13.9 further demonstrates that the RE-based metallic glasses with T_C lower than 200 K can only be used in the temperature region of liquid nitrogen or lower (Huo et al., 2013). Comparably, the Gd and Gd-based crystalline alloys with higher T_C meet the room-temperature refrigerant requirement. However, high material cost and poor corrosion resistance limit their massive productions. Consequently, Gd-based GdNi (Du et al., 2008), GdMn (Min et al., 2014), and GdCo (Liu et al., 2016) amorphous glasses with larger electrical resistivity together with higher thermal and corrosion resistance than the crystalline Gd have been developed. For the crystalline magnetic refrigerants, the $Gd_5Si_2Ge_2$ with giant and reversible

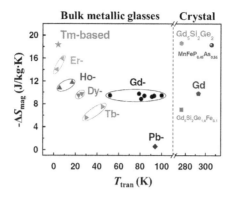

FIGURE 13.9 Comparison of the magnetic entropy changes $-\Delta S_{mag}$ under 5 T and transition temperature T_{tran} for bulk metallic glasses and crystalline magnetic refrigerants. Reprinted with permission from (Huo et al., 2013). Copyright (2013) Elsevier.

magnetocaloric effects in an alternating magnetic field has been widely investigated, which is originated from the first-order magnetic-structural transition from ferromagnetic orthorhombic to monoclinic at 276 K (Pecharsky and Gschneidner, 1997; Zhuang et al., 2006; Kou et al., 2020). The magnetic entropy change ΔS_{mag} of ~18.5 J/kg·K is almost twofold of that for the pure Gd, which is regarded as a milestone in the discovery of giant magnetocaloric materials with great application prospects.

13.3.2.2 Mn-Based Alloys

Since the T_C of many Mn-based compounds is near room temperature, the magnetocaloric properties can be improved by inducing the first-order phase transition. The MnAs is the first discovered Mn-based compound with large magnetocaloric effect. When demagnetized near the T_C of 318 K, the MnAs undergoes a first-order magnetic and structural transformation from ferromagnetic NiAs-type to paramagnetic MnP-type (Habiba et al., 2020). The concurrent magnetic and structural transitions contribute to the giant ΔS_{mag} of 32 J/kg·K for a field change of 0–5 T, which is about four times that of the pure Gd (Wada and Tanabe, 2001; Wada et al., 2003). However, since such first-order magnetic transition is accompanied with large thermal hysteresis, Sb substitution for the toxic As has been proposed to reduce the hysteresis with retained large magnetocaloric effect and lowered T_C (Wada and Tanabe, 2001). The MnFePAs compound with the Fe$_2$P-type crystal structure exhibits satisfactory room-temperature magnetocaloric performance (Tegus et al., 2013). After doping a small amount of interstitial B, not only the near room-temperature T_C (295 K) and high ΔS_{mag} (15.2 J/kg·K) can be obtained, the thermal hysteresis can also be effectively reduced. By Si substitution for As, the As-free MnFePSi compound possesses high ΔS_{mag} of 31 J/kg·K under a field change of 0–5 T at 304 K, but with the drawback of large thermal hysteresis (Dung et al., 2012; Cam Thanh et al., 2008).

13.3.2.3 Heusler Alloys

Heusler alloys refer to a large family of intermetallic compounds with the chemical formula of X_2YZ, where X and Y are transitional elements, Z is semiconductor or nonmagnetic element from Group IIIA, IVA or VA (Planes et al., 2009). In the magnetocaloric field, the Ni_2MnZ-based (Z = Ga, Sn, Sb, and In) Heusler alloys are the research focus. The NiMnGa Heusler alloys exhibit martensitic to austenitic structural transition between 175–220 K, and simultaneous ferromagnetic to paramagnetic transition between 315–380 K, which is strongly dependent on the ratios of Ni, Mn and Ga (Gschneidner et al., 2005). Albertini et al. have reported a maximum $|\Delta S_M|$ value of ~15 J/kg·K in $Ni_{2+x}Mn_{1-x}Ga$ compounds (x = 0.18 and 0.19), which is comparable to those of the $Gd_5Si_2Ge_2$ and $MnFeP_{0.45}As_{0.55}$ (Albertini et al., 2004). The Ga-free $Ni_{50}Mn_{50-x}Z_x$ (Z = Sn, Sb, In) Heusler alloys are promising candidates in commercial cooling devices (Kainuma et al., 2008), due to their austenite-martensite phase transition near room temperature, adjustable working temperature, low cost, non-toxicity and large ΔS_M (Ghosh and Ghosh, 2020; Devi et al., 2019). Qu et al. have reported that the 1 at% Ti-doped NiCoMnSn alloy exhibits a large reversible ΔS_M of 18.7 J/kg·K at 5 T, accompanied with satisfactory mechanical properties during thermal cycling (Qu et al., 2017).

13.3.2.4 LaFeSi Alloys

In 2001, Hu et al. firstly reported that the $NaZn_{13}$-type $LaFe_{11.4}Si_{1.6}$ compound possesses a large ΔS_M near its T_C of 208 K, which is originated from a negative lattice expansion from the low-temperature ferromagnetic state (lattice constant a = 11.52 Å) to the high-temperature paramagnetic state (lattice constant a = 11.48 Å) (Hu et al., 2001). With the applied magnetic field of 0–5 T, the maximum ΔS_M reaches 19.4 J/kg·K, which is correlated with the field-induced itinerant-electron metamagnetic transition in the vicinity of T_C (Scheibel et al., 2018; Hu et al., 2018). The substitution of Co for Fe or interstitial atom H have been reported to adjust the phase transition to be around the ambient temperature with maintained large magnetocaloric effect (Fujita et al., 2003; Hu et al., 2002; Zhang et al., 2016). Given the availability of the elemental constituents and satisfactory magnetocaloric performance, the LaFeSi alloy is considered to be one of the most promising systems for refrigeration around room temperature.

During the past century, continual endeavors of researchers and industrialists have led to the developments of important magnetocaloric materials. Deeper understandings on the fundamental mechanisms such as the hysteresis sources and the microscopic origins remain important tasks toward widespread applications in potential cooling devices.

REFERENCES

Abadian, S., G. Magno, V. Yam, and B. Dagens. 2021. Broad-band plasmonic isolator compatible with low-gyrotropy magneto-optical material. *Optics Express* 29(3):4091–104.

Akhtar, M. N., A. B. Sulong, M. A. Khan, M. Ahmad, G. Murtaza, M. R. Raza, R. Raza, M. Saleem, and M. Kashif. 2016. Structural and magnetic properties of yttrium iron garnet (YIG) and yttrium aluminum iron garnet (YAIG) nanoferrites prepared by microemulsion method. *Journal of Magnetism and Magnetic Materials* 401:425–31.

Albertini, F., F. Canepa, S. Cirafici, E. A. Franceschi, M. Napoletano, A. Paoluzi, L. Pareti, and M. Solzi. 2004. Composition dependence of magnetic and magnetothermal properties of Ni-Mn-Ga shape memory alloys. *Journal of Magnetism and Magnetic Materials* 272:2111–2.

Aoshima, K., R. Ebisawa, N. Funabashi, K. Kuga, and K. Machida. 2018. Current-induced domain-wall motion in patterned nanowires with various Gd-Fe compositions for magneto-optical light modulator applications. *Japanese Journal of Applied Physics* 57(9S2):09TC03.

Apicella, V., C. S. Clemente, D. Davino, D. Leone, and C. Visone. 2019. Review of modeling and control of magnetostrictive actuators. *Actuators* 8(2):45.

Arthur, E., and M. Clark. 2002. Magnetostrictive properties of Galfenol alloys under compressive stress. *Materials Transactions* 43(5):881–6.

Aung, Y. L., and A. Ikesue. 2017. Development of optical grade $(Tb_xY_{1-x})_3Al_5O_{12}$ ceramics as Faraday rotator material. *Journal of the American Ceramic Society* 100(9):4081–7.

Bodnar, W., M. Szklarska-Łukasik, P. Stoch, P. Zachariasz, and J. Pszczoła. 2010. Mossbauer effect studies of $Tb_{0.27}Dy_{0.73}(Fe_{1-x}Co_x)_2$ intermetallics at 295 K. *Pramana* 75(3):537–48.

Boisse, J., H. Zapolsky, and A. G. Khachaturyan. 2011. Atomic-scale modeling of nanostructure formation in Fe-Ga alloys with giant magnetostriction: cascade ordering and decomposition. *Acta Materialia* 59(7):2656–68.

Brown, G. V. 1976. Magnetic heat pumping near room temperature. *Journal of Applied Physics* 47(8):3673–80.

Busbridge, C. S., and A. R. Piercy. 1995. Magnetomechanical properties and anisotropy compensation in quaternary rare earth-iron materials of the type $Tb_xDy_yHo_zFe_2$. *IEEE Transactions on Magnetics* 31(6):4044–6.

Cam Thanh, D. T., E. Brück, N. T. Trung, J. C. P. Klaasse, K. H. J. Buschow, Z. Q. Ou, O. Tegus, and L. Caron. 2008. Structure, magnetism, and magnetocaloric properties of $MnFeP_{1-x}Si_x$ compounds. *Journal of Applied Physics* 103(7):07B318.

Chaudhary, V., X. Chen, and R. V. Ramanujan. 2019. Iron and manganese based magnetocaloric materials for near room temperature thermal management. *Progress in Materials Science* 100:64–98.

Chen, C., S. M. Zhou, H. Lin, and Q. Yi. 2012. Fabrication and performance optimization of the magneto-optical $(Tb_{1-x}R_x)_3Al_5O_{12}$ (R=Y, Ce) transparent ceramics. *Applied Physics Letters* 101(13):131908.

Chen, C., X. L. Li, Y. Feng, H. Lin, X. Z. Yi, Y. R. Tang, S. Zhang, and S. M. Zhou. 2015a. Optimization of CeO_2 as sintering aid for $Tb_3Al_5O_{12}$ Faraday magneto-optical transparent ceramics. *Journal of Materials Science* 50(6):2517–21.

Chen, J., H. Lin, D. M. Hao, Y. R. Tang, X. Z. Yi, Y. A. Zhao, and S. M. Zhou. 2019. Exaggerated grain growth caused by ZrO_2-doping and its effect on the optical properties of $Tb_3Al_5O_{12}$ ceramics. *Scripta Materialia* 162:82–5.

Chen, Q. L., and Q. H. Ma. 2020. Mixed samarium valences effect in Faraday rotation glasses: Structure, optical, magnetic and magneto-optical properties. *Journal of Non-Crystalline Solids* 530:119803.

Chen, Z., Y. Hang, L. Yang, J. Wang, X. Y. Wang, P. X. Zhang, J. Q. Hong, C. J. Shi, and Y. Q. Wang. 2015b. Great enhancement of Faraday effect by Pr doping terbium gallium garnet, a highly transparent VI-IR Faraday rotator. *Materials Letters* 145:171–3.

Clark, A. E., and H. S. Belson. 1972. Giant room-temperature magnetostrictions in $TbFe_2$ and $DyFe_2$. *Physical Review B* 5(9):3642–4.

Clark, A. E., B. F. DeSavage, and R. Bozorth. 1965. Anomalous thermal expansion and magnetostriction of single-crystal dysprosium. *Physical Review* 138(1A):216–24.

Clark, A. E., H. S. Belson, and N. Tamagawa. 1972. Huge magnetocrystalline anisotropy in cubic rare earth-Fe_2 compounds. *Physics Letters A* 42(2):160–162.

Clark, A. E., K. B. Hathaway, M. Wun-Fogle, J. B. Restorff, T. A. Lograsso, V. M. Keppens, G. Petculescu, and R. A. Taylor. 2003. Extraordinary magnetoelasticity and lattice softening in bcc Fe-Ga alloys. *Journal of Applied Physics* 93(10):8621–3.

Clark, A. E., M. Wun-Gogle, J. R. Restorff, T. A. Lograsso, and J. R. Cullen. 2001. Effect of quenching on the magnetostriction on $Fe_{1-x}Ga_x$ (0.13<x<0.21). *IEEE Transactions on Magnetics* 37(4):2678–80.

Debye, P. 1926. Einige Bemerkungen zur Magnetisierung bei tiefer Temperatur. *Annalen der Physik* 386:1154–60.

Deng, Z. X., and M. J. Dapino. 2017. Review of magnetostrictive vibration energy harvesters. *Smart Materials and Structures* 26(10):103001.

Devi, P., C. Salazar Mejía, M. Ghorbani Zavareh, K. K. Dubey, Y. Pallavi Kushwaha, C. Skourski, M. N. Felser, and S. Singh. 2019. Improved magnetostructural and magnetocaloric reversibility in magnetic Ni-Mn-In shape-memory Heusler alloy by optimizing the geometric compatibility condition. *Physical Review Materials* 3(6):062401(R).

Dillon Jr, J. 1958. Optical properties of several ferrimagnetic garnets. *Journal of Applied Physics* 29(3):539–41.

Du, J., Q. Zheng, Y. B. Li, Q. Zhang, D. Li, and Z. D. Zhang. 2008. Large magnetocaloric effect and enhanced magnetic refrigeration in ternary Gd-based bulk metallic glasses. *Journal of Applied Physics* 103(2):023918.

Dung, N. H., L. Zhang, Z. Q. Ou, and E. Brück. 2012. Magnetoelastic coupling and magnetocaloric effect in hexagonal Mn-Fe-P-Si compounds. *Scripta Materialia* 67(12):975–8.

Elisa, M., R. Iordanescu, C. Vasiliu, B. A. Sava, L. Boroica, M. Valeanu, V. Kuncser, A. C. Galca, A. Volceanov, M. Eftimie, A. Melinescu, and A. Beldiceanu. 2017. Magnetic and magneto-optical properties of Bi and Pb-containing aluminophosphate glass. *Journal of Non-Crystalline Solids* 465:55–8.

Faraday, M. 1846. I. Experimental researches in electricity.—Nineteenth series. *Philosophical Transactions of the Royal Society of London* 136:1–20.

Fujita, A., S. Fujieda, Y. Hasegawa, and K. Fukamichi. 2003. Itinerant-electron metamagnetic transition and large magnetocaloric effects in La(Fe$_x$Si$_{1-x}$)$_{13}$ compounds and their hydrides. *Physics Review B* 67(10):104416

Furuya, A., A. Sasaki, H. Morimura, O. Kagami, and T. Tanabe. 2016. Magnetooptical and crystalline properties of sputtered garnet ferrite film on spinel ferrite buffer layer. *Japanese Journal of Applied Physics* 55(9S):09SD01.

Geho, M., T. Sekijima, and T. Fujii. 2004. Growth of terbium aluminum garnet (Tb$_3$Al$_5$O$_{12}$; TAG) single crystals by the hybrid laser floating zone machine. *Journal of Crystal Growth* 267(1–2):188–93.

Geho, M., T. Sekijima, and T. Fujii. 2005. Growth mechanism of incongruently melting terbium aluminum garnet (Tb$_3$Al$_5$O$_{12}$; TAG) single crystals by laser FZ method. *Journal of Crystal Growth* 275(1–2):663–7.

Ghosh, S., and S. Ghosh. 2020. Cosubstitution in Ni-Mn-Sb Heusler compounds: realization of room-temperature reversible magnetocaloric effect driven by second-order magnetic transition. *Physical Review Materials* 4(2):025401.

Giauque, W. F. 1927. A thermodynamic treatment of certain magnetic effects. A proposed method of producing temperatures considerably below 1° absolute. *Journal of the American Chemical Society* 49(8):1864–70.

Giauque, W. F., and D. P. MacDougall. 1933. Attainment of temperatures below 1° absolute by demagnetization of Gd$_2$(SO$_4$)$_3$·8H$_2$O. *Physical Review* 43(9):768.

Golis, E. 2016. The effect of Nd^{3+} impurities on the magneto-optical properties of TeO$_2$-P$_2$O$_5$-ZnO-LiNbO$_3$ tellurite glass. *RSC Advances* 6(27):22370–3.

Gschneidner Jr, K. A., V. K. Pecharsky, and A. O. Tsokol. 2005. Recent developments in magnetocaloric materials. *Reports on Progress in Physics* 68(6):1479–539.

Habiba, U., K. S. Khattak, S. Ali, and Z. H. Khan. 2020. MnAs and MnFeP$_{1-x}$As$_{x}$-based magnetic refrigerants: a review. *Materials Research Express* 7(4):046106.

Haddadpour, A., V. F. Nezhad, Z. Yu, and G. Veronis. 2016. Highly compact magneto-optical switches for metal-dielectric-metal plasmonic waveguides. *Optics Letters* 41(18):4340–3.

Hao, D. M., J. Chen, G. Ao, Y. N. Tian, Y. R. Tang, X. Z. Yi, and S. M. Zhou. 2019. Fabrication and performance investigation of Thulium-doped TAG transparent ceramics with high magneto-optical properties. *Optical Materials* 94:311–5.

Hao, D. M., X. C. Shao, Y. R. Tang, X. Z. Yi, J. Chen, and S. M. Zhou. 2018. Effect of Si^{4+} doping on the microstructure and magneto-optical properties of TAG transparent ceramics. *Optical Materials* 77:253–7.

Hayakawa, T., M. Nogami, N. Nishi, and N. Sawanobori. 2002. Faraday rotation effect of highly Tb$_2$O$_3$/Dy$_2$O$_3$-concentrated B$_2$O$_3$-Ga$_2$O$_3$-SiO$_2$-P$_2$O$_5$ glasses. *Chemistry of Materials* 14(8):3223–5.

He, Y. K., C. B. Jiang, W. Wu, B. Wang, H. P. Duan, H. Wang, T. L. Zhang, J. M. Wang, J. H. Liu, Z. L. Zhang, P. Stamenov, J. M. D. Coey, and H. B. Xu. 2016. Giant heterogeneous magnetostriction in Fe-Ga alloys: effect of trace element doping. *Acta Materialia* 109:177–86.

Higashida, R., Funabashi, N., and Aoshima, K. I. 2020. Diffraction of light using high-density magneto-optical light modulator array. *Optical Engineering* 59(6):1–13.

Hu, F. X., B. G. Shen, J. R. Sun, G. J. Wang, and Z. H. Cheng. 2002. Very large magnetic entropy change near room temperature in LaFe$_{11.2}$Co$_{0.7}$Si$_{1.1}$. *Applied Physics Letters* 80(5):826–8.

Hu, F. X., B. G. Shen, J. R. Sun, Z. H. Cheng, G. H. Rao, and X. X. Zhang. 2001. Influence of negative lattice expansion and metamagnetic transition on magnetic entropy change in the compound LaFe$_{11.4}$Si$_{1.6}$. *Applied Physics Letters* 78(23):3675–7.

Hu, F., F. Shen, J. Hao, Y. Liu, J. Wang, J. Sun, and B. Shen. 2018. Negative thermal expansion in the materials with giant magnetocaloric effect. *Frontiers in Chemistry* 6:438.

Huo, J. T., D. Q. Zhao, H. Y. Bai, E. Axinte, and W. H. Wang. 2013. Giant magnetocaloric effect in Tm-based bulk metallic glasses. *Journal of Non-Crystalline Solids* 359:1–4.

Ishibashi, T., A. Mizusawa, N. Togashi, T. Mogi, M. Houchido, and K. Sato. 2005. (Re,Bi)$_3$(Fe,Ga)$_5$O$_{12}$ (ReY, Gd and Nd) thin films grown by MOD method. *Journal of Crystal Growth* 275(1–2):2427–31.

Ishida, E., K. Miura, Y. Shoji, T. Mizumoto, N. Nishiyama, and S. Arai. 2016. Magneto-optical switch with amorphous silicon waveguides on magneto-optical garnet. *Japanese Journal of Applied Physics* 55(8):088002.

Jin, W. Z., J. X. Ding, L. Guo, Q. Gu, C. Li, L. B. Su, A. H. Wu, and F. M. Zeng. 2018. Growth and performance research of $Tb_3Ga_5O_{12}$ magneto-optical crystal. *Journal of Crystal Growth* 484:17–20.

Kainuma, R., K. Oikawa, W. Ito, Y. Sutou, T. Kanomata, and K. Ishida. 2008. Metamagnetic shape memory effect in NiMn-based Heusler-type alloys. *Journal of Materials Chemistry* 18(16):1837.

Khachaturyan, A. G., and Viehland, D. 2007. Structurally heterogeneous model of extrinsic magnetostriction for Fe-Ga and similar magnetic alloys: part II. Giant magnetostriction and elastic softening. *Metallurgical and Materials Transactions A* 38:2308–28.

Kitanovski, A. 2020. Energy applications of magnetocaloric materials. *Advanced Energy Materials* 10(10):1903741.

Kou, R. H., J. R. Gao, Z. H. Nie, Y. D. Wang, D. E. Brown, and Y. Ren. 2020. Magnetic transitions and magnetocaloric effect of $Gd_4Nd_1Si_2Ge_2$. *Journal of Alloys and Compounds* 826:154117.

Law, J. Y. 2012. *The magnetocaloric effect of iron-based soft magnetic alloys.* Singapore: Nanyang Technological University.

Lee, W. E. 1955. Magnetostriction and magnetomechanical effects. *Reports on Progress in Physics* 18(1):184–229.

Legvold, S., J. Alstad, and J. Rhyne. 1963. Giant magnetostriction in dysprosium and holmium single crystals. *Physical Review Letters* 10(12):509–11.

Liu, G. L., D. Q. Zhao, H. Y. Bai, W. H. Wang, and M. X. Pan. 2016. Room temperature table-like magnetocaloric effect in amorphous $Gd_{50}Co_{45}Fe_5$ ribbon. *Journal of Physics D: Applied Physics* 49(5):055004.

Liu, H., H. Y. Lin, J. J. Ruan, D. Sun, and L. M. Song. 2020. A special intracavity power-modulator using the TGG magneto-optical effect. *Optik* 212:164739.

Liu, H., J. Shen, X. Liu, X. Chen, X. Hu, and N. Zhuang. 2021. Edge-defined film-fed growth of incongruent-melting Ce,Ga:GIG crystal with high magneto-optical performance. *Journal of Alloys and Compounds* 888:161456.

Liu, J. H., C. B. Jiang, and H. B. Xu. 2012. Giant magnetostrictive materials. *Science China Technological Sciences* 55(5):1319–26.

Lomako, I. D. 2013. Determination of the concentration of conduction electrons in $Y_3Fe_5O_{12}$ garnet crystals. *Crystallography Reports* 58(4):634–40.

Lv, X. K., S. W. Or, W. Liu, X. H. Liu, and Z. D. Zhang. 2009. Structural, magnetic, and magnetostrictive properties of Laves $(Tb_{0.3}Dy_{0.7})_{1-x}Pr_xFe_{1.55}$ ($0 \leq x \leq 0.4$) alloys. *Journal of Alloys and Compounds* 476(1–2):24–7.

Malakhovskii, A. V., I. S. Edelman, Y. Radzyner, Y. Yeshurun, A. M. Potseluyko, T. V. Zarubina, A. V. Zamkov, and A. I. Zaitzev. 2003. Magnetic and magneto-optical properties of oxide glasses containing Pr^{3+}, Dy^{3+} and Nd^{3+} ions. *Journal of Magnetism and Magnetic Materials* 263(1–2):161–72.

Man, P. W., F. K. Ma, T. Xie, J. X. Ding, A. H. Wu, L. B. Su, H. Y. Li, and G. H. Ren. 2017. Magneto-optical property of terbium-lutetium-aluminum garnet crystals. *Optical Materials* 66:207–10.

McCord, J. 2015. Progress in magnetic domain observation by advanced magneto-optical microscopy. *Journal of Physics D: Applied Physics* 48(33):333001.

Min, J. X., X. C. Zhong, Z. W. Liu, Z. G. Zheng, and D. C. Zeng. 2014. Magnetic properties and magnetocaloric effects of Gd-Mn-Si ribbons in amorphous and crystalline states. *Journal of Alloys and Compounds* 606:50–4.

Muñoz-Noval, A., S. Fin, E. Salas-Colera, D. Bisero, and R. Ranchal. 2018. The role of surface to bulk ratio on the development of magnetic anisotropy in high Ga content $Fe_{100-x}Ga_x$ thin films. *Journal of Alloys and Compounds* 745:413–20.

Murakami, M., and M. Birukawa. 2003. Effect of RE-TM underlayer on the microstructure of TbFeCo memory layer for high-density magneto-optical recording. *IEEE Transactions on Magnetics* 39(5):3178–80.

Nakamura, Y., H. Takagi, P. B. Lim, and M. Inoue. 2014. Magnetic volumetric hologram memory with magnetic garnet. *Optics Express* 22(13)16439–44.

Narita, F., Z. Wang, H. Kurita, Z. Li, Y. Shi, Y. Jia, and C. Soutis. 2021. A review of piezoelectric and magnetostrictive biosensor materials for detection of COVID-19 and other viruses. *Advanced Materials* 33(1):2005448.

Narita, F., and M. Fox. 2018. A review on piezoelectric, magnetostrictive, and magnetoelectric materials and device technologies for energy harvesting applications. *Advanced Engineering Materials* 20(5):1700743.

Nolting, A. E., and Eric Summers. 2015. Tensile properties of binary and alloyed Galfenol. *Journal of Materials Science* 50(15):5136–44.

Okazeri, K., K. Muraoka, Y. Shoji, S. Nakagawa, N. Nishiyama, S. Arai, and T. Mizumoto. 2018. Self-holding magneto-optical switch integrated with thin-film magnet. *IEEE Photonics Technology Letters* 30(4):371–4.

Pecharsky, V. K. and K. A. Gschneidner, Jr. 1997. Giant Magnetocaloric Effect in Gd5(Si2Ge2). *Physics Review Letters* 78(23):4494–7.

Planes, A., L. Manosa, and M. Acet. 2009. Magnetocaloric effect and its relation to shape-memory properties in ferromagnetic Heusler alloys. *Journal of Physics: Condensed Matter* 21(23):233201.

Potseluyko, A., I. Edelman, A. Malakhovskii, Y. Yeshurun, T. Zarubina, A. Zamkov, and A. Zaitsev. 2003. RE containing glasses as effective magneto-optical materials for 200–400 nm range. *Microelectronic Engineering* 69(2–4):216–20.

Qiu, Z. Q., and S. D. Bader. 2000. Surface magneto-optic Kerr effect. *Review of Scientific Instruments* 71(3):1243–55.

Qu, Y. H., D. Y. Cong, X. M. Sun, Z. H. Nie, W. Y. Gui, R. G. Li, Y. Ren, and Y. D. Wang. 2017. Giant and reversible room-temperature magnetocaloric effect in Ti-doped Ni-Co-Mn-Sn magnetic shape memory alloys. *Acta Materialia* 134:236–48.

Scheibel, F., T. Gottschall, A. Taubel, M. Fries, K. P. Skokov, A. Terwey, W. Keune, K. Ollefs, H. Wende, M. Farle, M. Acet, O. Gutfleisch, and M. E. Gruner. 2018. Hysteresis design of magnetocaloric materials-from basic mechanisms to applications. *Energy Technology* 6(8):1397–428.

Shen, Y., J. M. Ge, R. T. Liu, W. Fan, and Y. X. Yang. 2018. Optimization of optical current sensor based on rare-earth magneto-optical glass. *Microwave and Optical Technology Letters* 61(2):490–7.

Srinivasan, K., C. Radu, D. Bilardello, P. Solheid, and B. J. H. Stadler. 2020. Interfacial and bulk magnetic properties of stoichiometric cerium doped terbium iron garnet polycrystalline thin films. *Advanced Functional Materials* 30(15):2000409.

Srisukhumbowornchai, N., and S. Guruswamy. 2001. Large magnetostriction in directionally solidified FeGa and FeGaAl alloys. *Journal of Applied Physics* 90(11):5680–8.

Srisukhumbowornchai, N., and S. Guruswamy. 2002. Influence of ordering on the magnetostriction of Fe-27.5 at.% Ga alloys. *Journal of Applied Physics* 92(9):5371–9.

Stadler, B. J. H., and T. Mizumoto. 2014. Integrated magneto-optical materials and isolators: a review. *IEEE Photonics Journal* 6(1):1–15.

Starobor, A., D. Zheleznov, O. Palashov, C. Chen, S. M. Zhou, and R. Yasuhara. 2014. Study of the properties and prospects of Ce:TAG and TGG magnetooptical ceramics for optical isolators for lasers with high average power. *Optical Materials Express* 4(10):2127.

Suzuki, F., F. Sato, H. Oshita, S. Yao, Y. Nakatsuka, and K. Tanaka. 2018. Large Faraday effect of borate glasses with high Tb^{3+} content prepared by containerless processing. *Optical Materials* 76:174–7.

Tanaka, K., K. Fujita, N. Soga, J. R. Qiu, and K. Hirao. 1997. Faraday effect of sodium borate glasses containing divalent europium ions. *Journal of Applied Physics* 82(2):840–4.

Tegus, O., L. H. Bao, and L. Song. 2013. Phase transitions and magnetocaloric effects in intermetallic compounds MnFeX (X=P, As, Si, Ge). *Chinese Physics B* 22(3):037506.

Tsunashima, S. 2001. Magneto-optical recording. *Journal of Physics D: Applied Physics* 34(17):87–102.

Vinogradov, S., A. Cobb, and G. Light. 2017. Review of magnetostrictive transducers (MsT) utilizing reversed Wiedemann effect. *AIP Conference Proceedings* 1806:020008.

Wada, H., and Y. Tanabe. 2001. Giant magnetocaloric effect of $MnAs_{1-x}Sb_x$. *Applied Physics Letters* 79(20):3302–4.

Wada, H., T. Morikawa, K. Taniguchi, T. Shibata, Y. Yamada, and Y. Akishige. 2003. Giant magnetocaloric effect of $MnAs_{1-x}Sb_x$ in the vicinity of first-order magnetic transition. *Physica B: Condensed Matter* 328(1–2):114–6.

Wang, N. J., Y. Liu, H. W. Zhang, X. Chen, and Y. X. Li. 2016. Fabrication, magnetostriction properties and applications of Tb-Dy-Fe alloys a review. *China foundry* 13(2):75–84.

Weiss, P., and A. Piccard. 1917. Le phénomène magnétocalorique. *Journal of Physics: Theories and Applications* 7(1):103–9.

Wu, Y. Y., Y. J. Chen, C. Z. Meng, H. Wang, X. Q. Ke, J. M. Wang, J. H. Liu, T. L. Zhang, R. H. Yu, J. M. D. Coey, C. B. Jiang, and H. B. Xu. 2019. Multiscale influence of trace Tb addition on the magnetostriction and ductility of <100> oriented directionally solidified Fe-Ga crystals. *Physical Review Materials* 3(3):033401.

Wu, R., Z. Yang, and J. Hong. 2003. First-principles determination of magnetic properties. *Journal of Physics Condensed Matter* 15(5):S587–98.

Wun-Fogle, M., J. B. Restorff, A. E. Clark, and J. F. Lindberg. 1998. Magnetization and magnetostriction of dendritic $Tb_xDy_yHo_zFe_{1.95}$ ($x + y + z = 1$) rods under compressive stress. *Journal of Applied Physics* 83(11):7279–81.

Xie, J. W., D. Fort, Y. J. Bi, and J. S. Abell. 2000. Microstructure and magnetostrictive properties of Tb-Dy-Fe (Al) alloys. *Journal of Applied Physics* 87(9):6295–7.

Xu, L. H., C. B. Jiang, C. G. Zhou, and H. B. Xu. 2008. Magnetostriction and corrosion resistance of $Tb_{0.3}Dy_{0.7}(Fe_{1-x}Si_x)_{1.95}$ alloys. *Journal of Alloys and Compounds* 455(1–2):203–6.

Xue, G. M., P. L. Zhang, X. Y. Li, Z. B. He, H. G. Wang, Y. N. Li, R. Ce, W. Zeng, and B. Li. 2018. A review of giant magnetostrictive injector (GMI). *Sensors and Actuators A: Physical* 273:159–81.

Yin, H. R., Y. Gao, Y. X. Gong, R. Buchanan, J. B. Song, and M. Y. Li. 2018. Wavelength dependence of Tb^{3+} doped magneto-optical glass Verdet constant. *Ceramics International* 44(9):10929–33.

Zhang, H., J. Liu, M. X. Zhang, Y. Y. Shao, Y. Li, and A. R. Yan. 2016. $LaFe_{11.6}Si_{1.4}H_y$/Sn magnetocaloric composites by hot pressing. *Scripta Materialia* 120:58–61.

Zhang, S. Y., P. Liu, X. D. Xu, and J. Zhang. 2018a. Effect of the MgO on microstructure and optical properties of TAG ($Tb_3Al_5O_{12}$) transparent ceramics using hot isostatic pressing. *Optical Materials* 80:7–11.

Zhang, Y. J., T. Y. Ma, M. Yan, J. Y. Jin, B. Wu, B. X. Peng, Y. S. Liu, M. Yue, and C. Y. Liu. 2018b. Post-sinter annealing influences on coercivity of multi-main-phase Nd-Ce-Fe-B magnets. *Acta Materialia* 146:97–105.

Zhuang, N. F., W. B. Chen, L. J. Shi, J. B. Nie, X. L. Hu, B. Zhao, S. K. Lin, and J. Z. Chen. 2013. A new technique to grow incongruent melting Ga:YIG crystals: the edge-defined film-fed growth method. *Journal of Applied Crystallography* 46(3):746–51.

Zhuang, Y. H., J. Q. Li, W. D. Huang, W. A. Sun, and W. Q. Ao. 2006. Giant magnetocaloric effect enhanced by Pb-doping in $Gd_5Si_2Ge_2$ compound. *Journal of Alloys and Compounds* 421(1–2):49–53.

FURTHER READING

Schmidt, M. A., L. Wondraczek, H. W. Lee, N. Granzow, N. Da, and P. S. J. Russell. 2011. Complex Faraday rotation in microstructured magneto-optical fiber waveguides. *Advanced Materials* 23:2681–8.

CHAPTER **14**

Magnetic Materials for Electromagnetic Wave Absorption

14.1 INTRODUCTION TO ELECTROMAGNETIC WAVE ABSORPTION

Magnetic materials may respond to alternative electromagnetic (EM) fields in various ways such as current induction and magnetic resonance. Such functions open up new applications of magnetic materials in combating EM radiations. With the continuous development of microelectronics, telecommunication and national defense, EM radiation has become a critical issue, which may not only interfere the normal function of electronics but is also harmful to human health (Adam et al., 2002). Consequently, it is important and urgent to develop materials to tackle the EM radiations.

Figure 14.1 shows interactions between the EM waves and the absorbing material. The incident EM waves are partially reflected on the surface of the material and partially enter into the material to be transformed into thermal energy via various mechanisms. The unconsumed waves then transmit through the material. For optimal absorption, two conditions are critical including the impedance matching between air and the absorber to allow maximum entrance of the incident waves, together with the attenuation capability to efficiently transform the EM wave energy into heat.

DOI: 10.1201/9781003216346-18

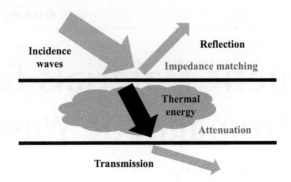

FIGURE 14.1 Interactions between the incident electromagnetic waves and the absorbing material.

14.1.1 Impedance Matching

If EM waves propagate in an infinite medium, the impedance Z can be expressed as

$$Z = \sqrt{\frac{\mu}{\varepsilon}} \qquad (14.1)$$

When EM waves incident onto the surface of a material with an input impedance Z_{in} from the free space with the impedance Z_0, the waves partially enter into the material, while the rest are reflected. The reflectivity R can be expressed as

$$R = \frac{Z_{in} - Z_0}{Z_{in} + Z_0} \qquad (14.2)$$

where $Z_0 = \sqrt{\frac{\mu_0}{\varepsilon_0}} = 1$ and $Z_{in} = Z_0 \sqrt{\frac{\mu_r}{\varepsilon_r}} \tanh\left[j\left(\frac{2\pi f d}{c}\right)\sqrt{\mu_r \varepsilon_r} \right]$. The μ_0 and ε_0 are the permeability and permittivity of the free space, while the μ_r, ε_r and d are the permeability, permittivity and thickness of the material, f is the EM wave frequency and c is the light velocity in vacuum. If $Z_{in} = Z_0 = 1$ and $R = 0$, optimal impedance matching is achieved between the absorbing material and the free space, resulting in complete entrance of the electromagnetic waves into the absorber without reflection. This is, however, an ideal case. During the design of high-performance wave absorbing materials, the EM parameters need to be adjusted to meet the impedance matching conditions as much as possible.

14.1.2 Attenuation Capability

The entered EM waves may be attenuated via various dielectric and magnetic loss mechanisms summarized in Table 14.1, details of which are introduced in the following subsections. The corresponding magnetic and dielectric loss tangent ($\tan\delta_e$ and $\tan\delta_m$) can be calculated based on

$$\tan\delta_e = \frac{\varepsilon''}{\varepsilon'} \quad (14.3)$$

$$\tan\delta_m = \frac{\mu''}{\mu'} \quad (14.4)$$

TABLE 14.1 Mechanisms for Dielectric and Magnetic Loss

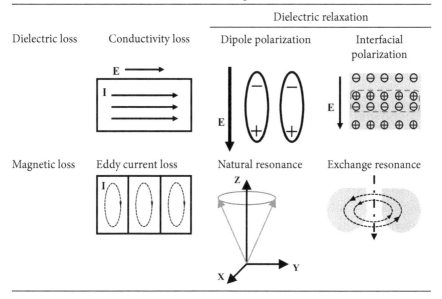

		Dielectric relaxation	
Dielectric loss	Conductivity loss	Dipole polarization	Interfacial polarization
Magnetic loss	Eddy current loss	Natural resonance	Exchange resonance

where ε'' and ε' are the imaginary and real part of the permittivity, while μ'' and μ' are the imaginary and real part of the permeability. Taking account of both dielectric and magnetic loss, the attenuation constant α reflecting the absorption capability can be calculated according to

$$\alpha = \frac{\sqrt{2}\pi f}{c} \times \sqrt{\left(\mu''\varepsilon'' - \mu'\varepsilon'\right) + \sqrt{\left(\mu''\varepsilon'' - \mu'\varepsilon'\right)^2 + \left(\mu'\varepsilon'' + \mu''\varepsilon'\right)^2}} \quad (14.5)$$

14.1.2.1 Dielectric Loss

The dielectric attenuation involves conductive loss and dielectric relaxation. The conductive loss originates from migration of the free electrons in the absorbing material under alternating electric field. Such electron migration induces current passing through the absorber to generate Joule heating for wave energy consumption. Absorbers with considerable conductive loss feature satisfactory electrical resistivity, such as carbon black (Liu et al., 2011; Pourya and Hasan, 2016), conductive polymers (Wu et al., 2017), graphene (Liu et al., 2019; Ye et al., 2018; Ma et al., 2018) and carbon nanotubes (Chen et al., 2014; Sun et al., 2014).

Relaxation occurs when a thermodynamic system deviates from the equilibrium under external interference and recovers after a period of time (Garrappa et al., 2016). Relative displacement of positive and negative charges results in the formation of dipolar moments. Such dipoles rearrange themselves along with alternating electric field directions, giving rise to dipole polarization. Also, interfacial polarization occurs at heterojunctions due to the movements of dipoles formed at the interfaces for EM wave attenuation. Typical absorbing materials involving dielectric relaxation include $BaCO_3$ (Kowsari and Karimzadeh, 2012), SiC (Zhao et al., 2021) and various heterojunctions (Lv et al., 2021).

14.1.2.2 Magnetic Loss

Magnetic loss mainly involves eddy current loss, natural and exchange resonance at the GHz frequencies. When a magnetic absorber is under alternating magnetic field, eddy current is generated due to electromagnetic induction for EM wave energy conversion. With dominating eddy current effect, the coefficient $C_0 = \mu''(\mu')^{-2}f^{-1}$ should be constant in spite of frequency variation (Wu et al., 2021). Magnetization of magnetic materials tends to occur along the direction with the lowest energy of the equivalent magnetic anisotropic field H_A. When the magnetic field direction of the EM waves is inconsistent with that of the easy magnetization, damping vibration takes place for the magnetic moments, i.e. Larmor precession to reduce energy. If the frequency of applied alternating magnetic field is equivalent to that of the damping vibration, natural resonance occurs. The frequency of natural resonance is associated to H_A based on $f_r = \gamma_0 H_A / 2\pi$ where the H_A mainly depends on crystalline anisotropy and shape anisotropy (Ma et al., 2010). Exchange interaction between adjacent magnetic moments within lattice is almost negligible for large grains/particles due to dominance of large magnetostatic energy directly proportional to the grain/particle volume. However, when the grain/particle size is sufficiently small to cause comparable exchange energy with the magnetostatic energy, exchange resonance occurs as another resonance mode, the frequency of which is related to grain/nanoparticle size R and intrinsic saturation magnetization M_s according to (Aharoni, 1991)

$$f = \frac{\gamma_0}{2\pi}\left(\frac{C\mu_{kn}^2}{R^2 M_s} + H_0 - NM_s + \frac{2K_1}{M_s}\right) \quad (14.6)$$

where f is the frequency of the exchange resonance, γ_0 is the gyromagnetic ratio, C is the exchange constant, μ_{kn} is the root of spherical Bessel function, H_0 is the applied magnetic field, N is the demagnetization factor and K_1 is the anisotropy constant.

Based on the loss mechanisms discussed above, the absorbers can be grouped into dielectric and magnetic ones, among which magnetic materials have attracted intensive interest due to their simultaneous magnetic and dielectric performance. Ferrites and magnetic alloys as two important categories in the development of advanced magnetic absorbers will be described in Section 14.2.

14.1.3 Evaluation of the Absorption Performance

The absorption performance is typically evaluated by reflection loss (RL, dB) based on transmission line theory as follows (Kwon et al., 1994)

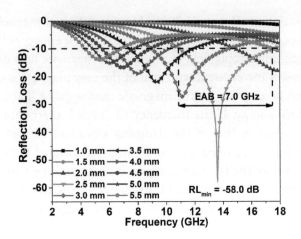

FIGURE 14.2 Typical reflection loss (RL) curves for FeNi$_3$/TiO$_2$ interconnected carbon fibers. Reprinted with permission from (Wang et al. 2021). Copyright (2021) American Chemical Society.

$$RL = 20\log\frac{|Z_{in} - Z_0|}{|Z_{in} + Z_0|} \quad (14.7)$$

where Z_0 is the impedance of free space and $Z_{in} = Z_0\sqrt{\frac{\mu_r}{\varepsilon_r}}\tanh\left[j\left(\frac{2\pi fd}{c}\right)\sqrt{\mu_r\varepsilon_r}\right]$ is the normalized input impedance as in Equation (14.2). Figure 14.2 illustrates typical graph of calculated RL at specific absorber thickness for FeNi$_3$/TiO$_2$ interconnected carbon fibers as an example (Wang et al., 2021). The minimum RL (RL$_{min}$) represents the absorption strength of the material, where RL = −10.0 dB, −20.0dB and −30.0 dB indicates 90%, 99% and 99.9% absorption of the electromagnetic waves at the corresponding matching thickness. The effective absorbing bandwidth (EAB) represents the frequency range where more than 90% (RL < −10 dB) of the electromagnetic waves are absorbed. Here the optimal performance (RL$_{min}$ = −58 dB and EAB = 7.0 GHz) can be achieved at a thickness of 2.5 mm. In addition to strong absorption, wide bandwidth and small thickness, the ideal electromagnetic wave absorber should be lightweight to meet the practical demand.

14.2 DEVELOPMENTS OF MAGNETIC ABSORBERS

14.2.1 Ferrites for Electromagnetic Wave Absorption

As an important type of magnetic materials, ferrites including both the hexagonal M ferrites and the spinal ferrites, have been used for EM wave absorption. Enhancement of the magnetic loss for the ferrites usually involves adjustment of the intrinsic magnetic properties (Garg et al., 2021; Saini et al., 2021; Mosleh et al., 2016) such as saturation magnetization and magnetocrystalline anisotropy via ion substitution. In the study of polycrystalline $Ba_{1-x}Ce_xFe_{12}O_{19}$, initial enhancement in magnetization followed by decrement has been observed with increased x to 0.20 (Mosleh et al., 2016). Since each Ce^{3+} possesses an unpaired electron, the replacement of the non-magnetic Ba^{2+} with the Ce^{3+} results in enhanced magnetization. Further raised Ce content leads to decreased magnetization, which can be attributed to partial conversion of the Fe^{3+} at the octahedral sites to Fe^{2+} at the tetrahedral site to maintain the charge balance. The RL_{min} are −16.74 dB and −20.47 dB when $x = 0.15$ and $x = 0.20$, comparing with $RL_{min} = -2.00$ dB for the pristine sample without doping. Substitution of the Fe^{3+} has also been investigated as in the barium hexagonal ferrite $BaNd_xFe_{12-x}O_{19}$ (Shivanshu et al., 2020). The doping of Nd gives rise to a RL_{min} of −33.20 dB with an EAB of 2.9 GHz (8.2 GHz–11.1 GHz) when $x = 0.04$, compared with $RL_{min} = -3.10$ dB for the undoped ferrite. Appropriate substitution of the Fe^{3+} with Nd^{3+} induces change in the direction of magneto-crystalline anisotropy for increased coercivity and magnetic loss, while excessive addition of Nd leads to the formation of Nd_2O_3 phase for deteriorated performance.

Besides doping, ferrites have also been composited with dielectric components for improved EM wave absorption with multiple loss mechanisms. Such compositing can be achieved via both physical and chemical routes. For the physical method, high-energy ball milling has usually been applied, as in the investigation of mixed $SrFe_{12}O_{19}$ and Ti_3SiC_2 powders (Avesh et al., 2020). The Ti_3SiC_2 alone exhibits metallic feature with poor impedance matching ($RL_{min} = -2.52$ dB). The introduction of the ferrite component not only improves the impedance matching but also brings the magnetic loss for enhanced absorption ($RL_{min} = -39.67$ dB). Chemical methods have also been frequently used to fabricate composites containing ferrites. For instance, $Co_{0.5}Zn_{0.5}Fe_2O_4$ has been grown on the flaky graphite surfaces via co-precipitation (Huang et al., 2016). The proportion of the $Co_{0.5}Zn_{0.5}Fe_2O_4$ and graphite sheets can be tuned for improved impedance matching and attenuation capability, giving rise to a RL_{min} of −33.85 dB with a thickness of 2.5 mm, compared to RL_{min} of −13.7 dB for the $Co_{0.5}Zn_{0.5}Fe_2O_4$ alone. Similar enhancements have also been achieved for $CoFe_2O_4$/graphene composites via hydrothermal (Fu et al., 2013) and vapor diffusion (Fu et al., 2014), as well as Fe_3O_4/graphene by catalytic chemical vapor deposition (Jian et al., 2016).

14.2.2 Metallic Magnetic Composites for Electromagnetic Wave Absorption

Different from the ferrites, magnetic alloys exhibit higher saturation magnetization and permeability for enhanced magnetic loss. Their large electrical conductivity, however, leads to poor impedance matching for reflection of the incident waves. To solve such problem, magnetic alloys are usually composited with dielectric materials for tuned impedance together with synergistic magnetic and dielectric loss.

The fabrication of the composite usually involves multiple steps, either by coating metallic magnetic particles with dielectric shells (Yang et al., 2017; Wu et al., 2015; Lopez-Ortega et al., 2015; Zhu et al., 2010) or growing magnetic components on dielectric supports (Zhang et al., 2021). For the former, core-shell Ni@SiO_2 exhibits a RL_{min} of −40.0 dB and an EAB of 3.5 GHz (10.9–14.4 GHz) at a thickness of merely 1.5 mm compared with the RL_{min} of −5.2 dB obtained for the Ni core alone (Zhao et al., 2015). In the consideration of lightweight and multiple scattering, air gap has been generated in the core-shell structure in the formation of a yolk-shell morphology. Figure 14.3a illustrates the synthesis of the CoNi@void@TiO_2 by etching the mid-layered SiO_2 (Liu et al., 2016). While the RL_{min} of the CoNi and CoNi@SiO_2 are −22.1 dB and −23.8 dB, respectively, the CoNi@void@TiO_2 microspheres exhibit remarkably improved RL_{min} of −46.7 dB and an EAB over 6 GHz at a thickness of 2.5 mm.

Dielectric materials, such as graphene oxide (Wang et al., 2015; Zhang et al., 2018) and MoS_2 (Pan et al., 2018) with a large specific surface area have been used as the support to grow dispersive magnetic nanoparticles. For instance, magnetic Ni particles have been grown on MoS_2 by electroless plating with controllable size and uniform distribution. Combined magnetic and dielectric loss as well as strong interfacial polarization give rise to the optimal RL_{min} of −22.0 dB and EAB of 2.8 GHz at a thickness of

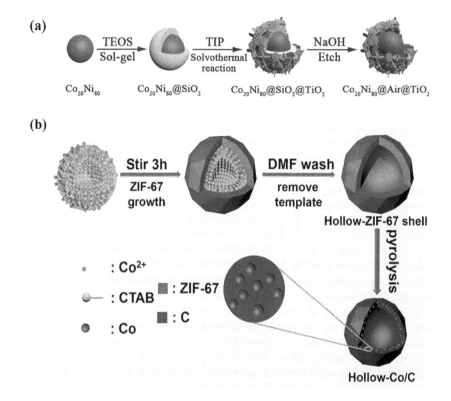

FIGURE 14.3 Schematic illustrations for the synthesis of (a) the CoNi@void@TiO$_2$ microspheres and (b) the MOF-derived Co/C hollow spheres. (a) Reprinted with permission from Liu et al. (2016). Copyright (2015) John Wiley and Sons. (b) Reprinted with permission from Li et al. (2018). Copyright (2018) American Chemical Society.

2.0 mm (Liu et al., 2020), while it is difficult to achieve RL$_{min}$ < 10 dB for the Ni alone (Gao et al., 2008).

Recently, metal-organic frames (MOFs) have been used as the precursors for one-step derivation of composites containing metallic magnetic nanoparticles and carbon (Ouyang et al., 2019; Zeng et al., 2021), or as the component for even more complex composites (Fan et al., 2009; Liu et al., 2005). Li et al. have used cetyltrimethylammonium bromide (CTAB) as the template to grow hollow ZIF-67 assemblies followed by pyrolysis in the fabrication of Co/C microspheres as shown in Figure 14.3b (Li et al., 2018). Characteristics including large surface area and pore volume have been maintained after pyrolysis in the formation of the Co/C for lightweight and enhanced wave-absorber interactions. As a result, the Co/C composite exhibits improved impedance matching and attenuation in the achievement of RL_{min} = −66.5 dB and EAB = 14.3 GHz. The MOFs can also be composited with other dielectric materials for further regulation of the electromagnetic parameters. Bimetallic Co/Zn-MOFs have been in situ grown on multi-walled carbon nanotubes (MWCNTs) which convert into Co/C/MWCNTs composites after pyrolysis (Shu et al., 2019). Not only functional groups on the surfaces of the MWCNTs serve as polarization centers, the heterogeneous interfaces among Co, C and MWCNTs could be considered as the capacitor-like structure for rapid EM wave consumption. Consequently, RL_{min} of −50.0 dB and EAB of 4.3 GHz have been achieved at a small thickness of 1.8 mm.

Besides the impedance match, another issue of the metallic magnetic materials lies in the Snoek limit given by (Snoek, 1948)

$$(\mu-1)f = \frac{2}{3}\gamma(4\pi)M_s \tag{14.8}$$

It indicates that the permeability μ and the cut-off frequency f cannot be simultaneously increased, since the saturation magnetization M_s is certain with determined material composition. Equation (14.8), however, can be transformed into the following when the anisotropy K is taken into consideration (Jonker et al. 1956)

$$(\mu-1)f = \frac{2}{3}\gamma(4\pi)KM_s \tag{14.9}$$

An effective approach to induce the K is via surface and shape anisotropy by the design of zero-, one- or two-dimensional (0D, 1D and 2D) materials. The 0D nanoparticles are beneficial for enhanced surface anisotropy and tunable magnetic loss. For the 0D magnetic nanoparticles, a major challenge remains to overcome their agglomeration. Zhou et al. have used self-assembled MoS_2 nanoflowers to provide numerous nucleation sites for the growth of dispersive FeCo nanoparticles (Zhou et al., 2018). The resultant FeCo approaching single domain size results in increased frequency of the natural resonance and excellent wave

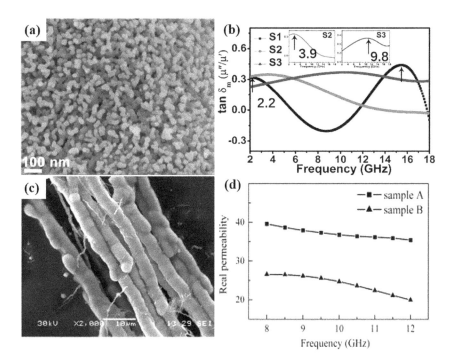

FIGURE 14.4 (a) SEM image showing the 0D heterogeneous FeCo/ZnO composite and (b) the increased natural resonant frequency from 2.2 GHz to 3.9 GHz and 9.8 GHz with enhancement confinement effect by the introduction of ZnO. Reprinted with permission from Zhou et al. (2019). Copyright (2019) Elsevier. (c) SEM image of the 1D $Fe_{55}Ni_{45}$ as sample A and (d) its improved permeability compared with sample B with weakened shape anisotropy. Reprinted with permission from Li et al. (2005) Copyright (2005) Elsevier.

absorption performance with RL_{min} of −64.6 dB and EAB of 7.2 GHz at a thickness of 2 mm. In another work, a confinement route has been developed to synthesize FeCo/ZnO via a facile glucose-assisted template method (Zhou et al., 2019). The growth of the FeCo nanoparticles (~20 nm) has been restricted by the adjacent ZnO particles as illustrated by the different contrast in Figure 14.4a. This results in increased natural resonance frequency from 2.2 GHz to 3.9 GHz and 9.8 GHz with increased ZnO content from 0.0 wt%, 10.0 wt% to 50.0 wt% as shown in Figure 14.4b.

Shape anisotropy can be effectively enhanced in 1D and 2D magnetic materials. For instance, 1D ferromagnetic $Fe_{55}Ni_{45}$ fibers have been prepared via thermal decomposition of polymetallic carbonyl with the assistance of magnetic field, during which the fiber dimensions can be tuned by the field strength (Li et al., 2005). Figure 14.4c shows the SEM image of the $Fe_{55}Ni_{45}$ fibers (sample A) with a diameter of 5 μm and an aspect ratio of around 60–80 prepared under the field of 2400 A/m. Higher permeability (Figure 14.4d) can be obtained for the fibers with smaller diameter and larger aspect ratio for sample A compared with those of the sample B, giving rise to improved RL_{min}.

2D materials are beneficial to introduce the plane anisotropy. Fe powders have been ball milled to form the flaky morphology. Prolonged milling time from 0.5 h to 4.0 h gives rise to enhanced plane anisotropy as well as increased μ' and μ'' from 2.2 to 6.6 and from 2.5 to 3.9, respectively (Kim et al., 2005). Similar enhancements have also been reported

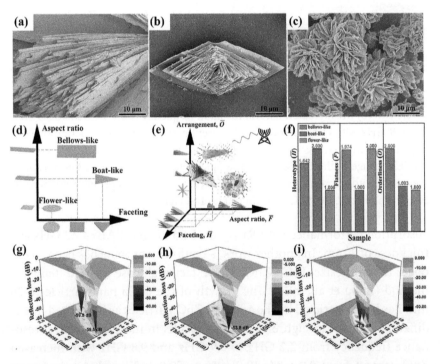

FIGURE 14.5 SEM images of the (a) bellows-, (b) boat-, and (c) flower-like MOFs. Anisotropy dimensions to describe the (d) building blocks and (e) their assemblies. (f) Comparisons between the anisotropy of the MOF assemblies. Reflection loss of the (g) bellows-, (h) boat-, and (i) flower-like Co/C composites. Reprinted with permission from Liu et al. (2021b). Copyright (2021) Elsevier.

for the FeSiAl flaky powders (Zhou et al., 2009). In fact, much effort has been devoted to fabricating magnetic absorbers with strong shape anisotropy but it remains challenging to precisely control the flatness and uniformity. Recently, 2D CoNi MOFs have been used as precursor to generate high-yield Co/Ni/C nanosheet arrays with large size, uniform shape and enhanced anisotropy, giving rise to satisfactory RL_{min} of −49.8 dB with a wide EAB of 7.6 GHz (Liu et al., 2021a). Hierarchical structures may also be formed via the assembling of 2D flakes via chemical routes, adding complexity to evaluate the anisotropy. MOFs with bellows-, boat- and flower-like morphologies (Figure 14.5a–c) have been fabricated, with their anisotropy investigated based on the dimensional facets, aspect ratio and arrangement of the assembled flakes as shown in Figure 14.5d–f (Liu et al., 2021b). Compared with the other two structures, the bellows-like MOF exhibits enhanced anisotropy for the optimal absorption performance with RL_{min} = −59.8 dB and EAB = 5.7 GHz (Figure 14.5g–i).

14.3 FUTURE WORK AND PERSPECTIVES

Magnetic materials possess the advantage of simultaneous dielectric and magnetic loss for EM wave absorption. While much effort has been devoted to developing ferrite- and magnetic-alloy-based absorbers with tuned composition and microstructure, the most prevalent absorbers remain in the powder form. With the emergency of foldable and wearable devices, flexible EM absorbers may become an important trend for research. The focuses may lie in (i) design strategy and fabrication technologies to incorporate the magnetic and dielectric components in flexible absorbers; and (ii) how the magnetic and dielectric loss mechanisms evolve when the absorbers are under deformation.

REFERENCES

Adam, J. D., L. E. Davis, G. F. Dionne, E. F. Schloemann, and S. N. Stitzer. 2002. Ferrite devices and materials. *IEEE Transactions on Microwave Theory and Techniques* 50:721–37.

Aharoni, A. 1991. Exchange resonance models in a ferromagnetic sphere. *Journal of Applied Physics* 69:7762–4.

Avesh, G., G. Shiwanshu, K. Neelam, D. Ashish, P. N. Eswara, and T. Sachin. 2020. Development of $SrFe_{12}O_{19}$-Ti_3SiC_2 composites for enhanced microwave absorption. *Journal of Electronic Materials* 49:2233–41.

Chen, M. D., X. H. Jie, and H. Y. Zhang. 2014. Simulation and calculation of the absorbing microwave properties of carbon nanotube composite coating. *Acta Physica Sinica* 63:066103.

Fan, X. A., J. G. Guan, W. Wang, and G. X. Tong. 2009. Morphology evolution, magnetic and microwave absorption properties of nano/submicrometre iron particles obtained at different reduced temperatures. *Journal of Physics D-Applied Physics* 42:075006.

Fu, M., Q. Z. Jiao, and Y. Zhao. 2013. In situ fabrication and characterization of cobalt ferrite nanorods/graphene composites. *Materials Characterization* 86:303–15.

Fu, M., Q. Z. Jiao, Y. Zhao, and H. S. Li. 2014. Vapor diffusion synthesis of $CoFe_2O_4$ hollow sphere/graphene composites as absorbing materials. *Journal of Materials Chemistry A* 2:735–44.

Gao, B., L. Qiao, J. B. Wang, Q. F. Liu, F. S. Li, J. Feng, and D. S. Xue. 2008. Microwave absorption properties of the Ni nanowires composite. *Journal of Physics D-Applied Physics* 41:235005.

Garg, A., S. Goel, A. K. Dixit, M. K. Pandey, N. Kumari, and S. Tyagi. 2021. Investigation on the effect of neodymium doping on the magnetic, dielectric and microwave absorption properties of strontium hexaferrite particles in X-band. *Materials Chemistry and Physics* 257:123771.

Garrappa, R., F. Mainardi, and G. Maione. 2016. Models of dielectric relaxation based on completely monotone functions. *Fractional Calculus and Applied Analysis* 19:1105–60.

Huang, X. G., J. Zhang, W. F. Rao, and T. Y. Sang. 2016. Tunable electromagnetic properties and enhanced microwave absorption ability of flaky graphite-cobalt zinc ferrite composites. *Journal of Alloys and Compounds* 662:409–14.

Jian, X., B. Wu, Y. F. Wei, S. X. Dou, X. L. Wang, W. D. He, and N. Mahmood. 2016. Facile synthesis of Fe_3O_4/GCs composites and their enhanced microwave absorption properties. *ACS Applied Materials & Interfaces* 8:6101–9.

Jonker, G. H., Wijn H. P. J. and P. B. Braun. 1956. *Philips Technical Revue* 18:145–148.

Kim, S. S., S. T. Kim, Y. C. Yoon, and K. S. Lee. 2005. Magnetic, dielectric, and microwave absorbing properties of iron particles dispersed in rubber matrix in gigahertz frequencies. *Journal of Applied Physics* 97:10F905.

Kowsari, E., and A. H. Karimzadeh. 2012. Using a chiral ionic liquid for morphological evolution of $BaCO_3$ and its radar absorbing properties as a dendritic nanofiller. *Materials Letters* 78:150–3.

Kwon, H. J., J. Y. Shin, and J. H. Oh. 1994. The microwave absorbing and resonance phenomena of Y-type hexagonal ferrite microwave absorbers. *Journal of Applied Physics* 75:6109.

Li, X. C., R. Z. Gong, Y. Nie, Z. S. Zhao, and H. H. He. 2005. Electromagnetic properties of $Fe_{55}Ni_{45}$ fiber fabricated by magnetic-field-induced thermal decomposition. *Materials Chemistry and Physics* 94:408–11.

Li, Z. N., X. J. Han, Y. Ma, D. W. Liu, Y. H. Wang, P. Xu, C. L. Li, and Y. C. Du. 2018. MOFs-derived hollow Co/C microspheres with enhanced microwave absorption performance. *ACS Sustainable Chemistry & Engineering* 6:8904–13.

Liu, G., J. Q. Tu, C. Wu, Y. J. Fu, C. H. Chu, Z. H. Zhu, X. H. Wang, and M. Yan. 2021a. High-yield two-dimensional metal–organic framework derivatives for wideband electromagnetic wave absorption. *ACS Applied Materials & Interfaces* 13:20459–66.

Liu, G., C. Wu, L. Hu, X. J. Hu, X. F. Zhang, J. Tang, H. F. Du, X. H. Wang, and M. Yan. 2021b. Anisotropy engineering of metal organic framework derivatives for effective electromagnetic wave absorption. *Carbon* 181:48–57.

Liu, J. R., M. Itoh, T. Horikawa, K. Machida, S. Sugimoto, and T. Maeda. 2005. Gigahertz range electromagnetic wave absorbers made of amorphous-carbon-based magnetic nanocomposites. *Journal of Applied Physics* 98:054305.

Liu, P. B., S. Gao, Y. Wang, Y. Huang, Y. Wang, and J. H. Luo. 2019. Core-shell CoNi@graphitic carbon decorated on B,N-codoped hollow carbon polyhedrons toward lightweight and high-efficiency microwave attenuation. *ACS Applied Materials & Interfaces* 11:25624–35.

Liu, Q. H., Q. Cao, H. Bi, C. Y. Liang, K. P. Yuan, W. She, Y. J. Yang, and R. C. Che. 2016. CoNi@SiO$_2$@TiO$_2$ and CoNi@Air@TiO$_2$ microspheres with strong wideband microwave absorption. *Advanced Materials* 28:486–90.

Liu, X. X., Z. Y. Zhang, and Y. P. Wu. 2011. Absorption properties of carbon black/silicon carbide microwave absorbers. *Composites Part B: Engineering* 42:326–9.

Liu, Yi, Chen Ji, Xiaolei Su, Xinhai He, Jie Xu, and Yunyu Li. 2020. Enhanced microwave absorption properties of flaky MoS$_2$ powders by decorating with Ni particles. *Journal of Magnetism and Magnetic Materials* 511:166961.

Lopez-Ortega, A., M. Estrader, G. Salazar-Alvarez, A. G. Roca, and J. Nogues. 2015. Applications of exchange coupled bi-magnetic hard/soft and soft/hard magnetic core/shell nanoparticles. *Physics Reports-Review Section of Physics Letters* 553:1–32.

Lv, H. B., C. Wu, J. Tang, H. F. Du, F. X. Qina, H. X. Peng, and M. Yan. 2021. Two-dimensional SnO/SnO$_2$ heterojunctions for electromagnetic wave absorption. *Chemical Engineering Journal* 411:128445.

Ma, F., Y. Qin, and Y. Z. Li. 2010. Enhanced microwave performance of cobalt nanoflakes with strong shape anisotropy. *Applied Physics Letters* 96:202507.

Ma, J. R., X. X. Wang, W. Q. Cao, C. Han, H. J. Yang, J. Yuan, and M. S. Cao. 2018. A facile fabrication and highly tunable microwave absorption of 3D flower-like Co$_3$O$_4$-rGO hybrid-architectures. *Chemical Engineering Journal* 339:487–98.

Mosleh, Z., P. Kameli, A. Poorbaferani, M. Ranjbar, and H. Salamati. 2016. Structural, magnetic and microwave absorption properties of Ce-doped barium hexaferrite. *Journal of Magnetism and Magnetic Materials* 397:101–7.

Ouyang, J., Z. L. He, Y. Zhang, H. M. Yang, and Q. H. Zhao. 2019. Trimetallic FeCoNi@C nanocomposite hollow spheres derived from metal-organic frameworks with superior electromagnetic wave absorption ability. *ACS Applied Materials & Interfaces* 11:39304–14.

Pan, J. J., X. Sun, T. Wang, Z. T. Zhu, Y. P. He, W. Xia, and J. P. He. 2018. Porous coin-like Fe@MoS$_2$ composite with optimized impedance matching for efficient microwave absorption. *Applied Surface Science* 457:271–9.

Pourya, M., and J. Hasan. 2016. Effect of carbon black content on the microwave absorbing properties of CB/epoxy composites. *Journal of Nanostructures* 6:140–8.

Saini, J., D. Yadav, V. Sharma, M. Sharma, and B. K. Kuanr. 2021. Effect of aliovalent substitution in $Y_{2.9}Bi_{0.1}Fe_5O_{12}$ magnetization dynamic study. *IEEE Transactions on Magnetics* 57:2200405.

Snoek, J.L.. 1948. *Physica* 4: 207–217.

Shivanshu, G., G. Avesh, G. Raju Kumar, D. Ashish, P. N. Eswara, and T. Sachin. 2020. Effect of neodymium doping on microwave absorption property of barium hexaferrite in X-band. *Materials Research Express* 7:016109.

Shu, R. W., W. J. Li, Y. Wu, J. B. Zhang, and G. Y. Zhang. 2019. Nitrogen-doped Co-C/MWCNTs nanocomposites derived from bimetallic metal-organic frameworks for electromagnetic wave absorption in the X-band. *Chemical Engineering Journal* 362:513–24.

Sun, H., R. C. Che, X. You, Y. S. Jiang, Z. B. Yang, J. Deng, L. B. Qiu, and H. S. Peng. 2014. Cross-stacking aligned carbon-nanotube films to tune microwave absorption frequencies and increase absorption intensities. *Advanced Materials* 26:8120–5.

Wang, J. P., J. Wang, R. X. Xu, Y. Sun, B. Zhang, W. H. Chen, T. Wang, and S. Yang. 2015. Enhanced microwave absorption properties of epoxy composites reinforced with $Fe_{50}Ni_{50}$-functionalized graphene. *Journal of Alloys and Compounds* 653:14–21.

Wang, Z. H., L. X. Yang, Y. Zhou, C. Xu, M. Yan, and C. Wu. 2021. NiFe LDH/MXene derivatives interconnected with carbon fabric for flexible electromagnetic wave absorption. *ACS Applied Materials & Interfaces* 13:16713–21.

Wu, S. Y., J. Yang, R. C. Yang, J. P. Zhu, and S. Liu. 2021. Preparation and properties of microwave-absorbing asphalt mixtures containing graphite and magnetite powder. *Journal of Testing and Evaluation* 49:573–89.

Wu, Y., Z. Y. Wang, X. Liu, X. Shen, Q. B. Zheng, Q. Xue, and J. K. Kim. 2017. Ultralight graphene foam/conductive polymer composites for exceptional electromagnetic interference shielding. *Acs Applied Materials & Interfaces* 9:9059–69.

Wu, Y. P., J. X. Cheng, W. P. Zhou, F. T. Zhao, W. W. Wen, J. H. Liu, and Y. X. Hu. 2015. Extinction performance of microwave by core-shell spherical particle. *IOP Conference Series: Materials Science and Engineering* 87:012001.

Yang, H. J., Q. L. Wang, L. Wang, Z. L. Zhang, Y. L. Li, and M. S. Cao. 2017. Improved dielectric properties and microwave absorbing properties of SiC Nanorods/Ni core-shell structure. *Functional Materials Letters* 10:1750069.

Ye, F., Q. Song, Z. C. Zhang, W. Li, S. Zhang, X. Yin, Y. Zhou, H. Tao, Y. Liu, L. Cheng, and L. Zhang. 2018. Direct growth of edge-rich graphene with tunable dielectric properties in porous Si_3N_4 ceramic for broadband high-performance microwave absorption. *Advanced Functional Materials* 28:1707205.

Zeng, Q. W., L. Wang, X. Li, W. B. You, J. Zhang, X. H. Liu, M. Wang, and R. C. Che. 2021. Double ligand MOF-derived pomegranate-like Ni@C microspheres as high-performance microwave absorber. *Applied Surface Science* 538:148051.

Zhang, B., J. Wang, H. Y. Tan, X. G. Su, S. Q. Huo, S. Yang, W. H. Chen, and J. P. Wang. 2018. Synthesis of Fe@Ni nanoparticles-modified graphene/epoxy composites with enhanced microwave absorption performance. *Journal of Materials Science* 29:3348–57.

Zhang, H. X., C. Shi, Z. R. Jia, X. H. Liu, B. H. Xu, D. D. Zhang, and G. L. Wu. 2021. FeNi nanoparticles embedded reduced graphene/nitrogen-doped carbon composites towards the ultra-wideband electromagnetic wave absorption. *Journal of Colloid and Interface Science* 584:382–94.

Zhao, B., G. Shao, B. B. Fan, W. Y. Zhao, and R. Zhang. 2015. Investigation of the electromagnetic absorption properties of $Ni@TiO_2$ and $Ni@SiO_2$ composite microspheres with core-shell structure. *Physical Chemistry Chemical Physics* 17:2531–9.

Zhao, Y. J., Y. N. Zhang, C. R. Yang, and L. F. Cheng. 2021. Ultralight and flexible SiC nanoparticle-decorated carbon nanofiber mats for broad-band microwave absorption. *Carbon* 171:474–83.

Zhen, X., Y. M. Song, J. Xiong, Z. B. Pan, X. Wang, L. Liu, R. Liu, H. W. Yang, and W. Lu. 2019. Enhanced electromagnetic wave absorption of nanoporous Fe_3O_4@carbon composites derived from metal-organic frameworks. *Carbon* 142:20–31.

Zhou, C. H., C. Wu, and M. Yan. 2018. Hierarchical FeCo@MoS_2 nanoflowers with strong electromagnetic wave absorption and broad bandwidth. *ACS Applied Nano Materials* 1:5179–87.

Zhou, C. H., C. Wu, and M. Yan. 2019. A versatile strategy towards magnetic/dielectric porous heterostructure with confinement effect for lightweight and broadband electromagnetic wave absorption. *Chemical Engineering Journal* 370:988–96.

Zhou, T. D., P. H. Zhou, D. F. Liang, and L. J. Deng. 2009. Structure and electromagnetic characteristics of flaky FeSiAl powders made by melt-quenching. *Journal of Alloys and Compounds* 484:545–9.

Zhu, C. L., M. L. Zhang, Y. J. Qiao, G. Xiao, F. Zhang, and Y. Y. Chen. 2010. Fe_3O_4/TiO_2 core/shell nanotubes: synthesis and magnetic and electromagnetic wave absorption characteristics. *Journal of Physical Chemistry C* 114:16229–35.

FURTHER READING

Wei, H., Z. Zhang, G. Hussain, L. Zhou, Q. Li, and K. Ostrikov. 2020. Techniques to enhance magnetic permeability in microwave absorbing materials. *Applied Materials Today* 19:100596.

CHAPTER 15

Magnetic Materials for Biomedicine, Catalysis and Others

15.1 MAGNETIC MATERIALS FOR BIOMEDICINE

Magnetic nanoparticles (MNPs) with low toxicity have been applied interdisciplinarily as in biomedicine. The MNPs not only possess high specific surface area and size effects similar to other nanoparticles, but also exhibit superparamagnetic properties with high saturation magnetization and low coercivity when the particle size is below the critical size. Due to the response of the MNPs to applied external field, they have been used in magnetic targeting, imaging diagnosis and tumor therapy, etc.

15.1.1 Magnetic Targeting

Due to pharmacokinetics, most of the drugs cannot accumulate at the target sites such as organs and cells, resulting in low efficacy and adverse effects on normal tissues. Guiding the drug to the target site becomes essential in clinical medicine. Since the movement of MNPs can be manipulated by external magnetic field, they have been used as drug carriers to accumulate at the target sites (Feng et al., 2018). For instance, doxorubicin (DOX) and polyethylene glycol (PEG) have been coupled on the surface of the superparamagnetic iron oxide (SPIO) to prolong its half-life in the blood circulation (Liang et al., 2016). Investigation on the anti-cancer efficiency of the SPIO-PEG-DOX reveals that the tumor size with the introduction of the magnetic field is significantly smaller than the counterpart without the magnetic field due to the accumulation of SPIO-PEG-DOX in the tumor tissues. Similar enhancement has also been demonstrated for Fe_3O_4/SiO_2 mesoporous particles coated with red cell membrane (Xuan et al., 2018).

Magnetic field targeting can be combined with photothermal therapy (PTT) and photodynamic therapy (PDT) for improved treating efficiency. MoS_2/Fe_3O_4 composites (MISOs) have been fabricated via a simple two-step hydrothermal route for magnetic targeting assisted PTT (Yu et al., 2015). Fluorescence images of in vitro cells indicate accumulation of the MISOs at the target site. Meanwhile, in vivo experiments in mice show rapid increment of the tumor temperature to 47 °C with the assistance of magnetic targeting, compared with 42 °C for the control group without magnetic field.

Although magnetic targeting is promising, a major issue is that the magnetic field rapidly weakens with increased distance, causing the induction limited to the surface area near the magnet and cannot be extended to the deep tissue. Recently, Liu et al. have proposed to use a dual-pole device, which is composed of two oppositely polarized magnets to generate a constant and uniform magnetic field gradient in-between (Liu et al., 2020). Compared with the single magnet configuration, permeability of the magnetic iron oxide nanoparticles at the tumor site have been increased by five times with tripled accumulation.

15.1.2 Magnetic Resonance Imaging

Magnetic resonance imaging (MRI) is one of the most commonly used imaging diagnostic techniques clinically. It utilizes contrast agent to enhance the contrast between normal and diseased tissues for improved sensitivity and accuracy of diagnosis. According to its effect on the proton relaxation time, the contrast agent can be divided into positive and negative ones, also known as the T_1- weighted and T_2- weighted contrast agent, the corresponding relaxation of which are denoted as r_1 and r_2. The T_1- and T_2- weighted contrast agent shortens the longitudinal and transverse relaxation time of hydrogen proton and enhance the brightness and darkness of magnetic resonance imaging, respectively.

The most widely used T_1 contrast agent is based on gadolinium with a series of gadopentetate chelates developed since the 1980s (Fraum et al., 2017). In recent years, however, studies have shown that the Gd-based contrast agents tend to release free Gd^{3+} in the metabolic process and induce nephrogenic systemic fibrosis (NSF) (Edward et al., 2008). Consequently, developing alternative contrast agent with low toxicity is a future research direction.

Compared with the Gd-based contrast agents, Fe-based contrast agents have attracted much attention due to their safety and biocompatibility. Magnetic iron oxide nanoparticles (MIONs) are usually used as the T_2 contrast agent and combined with other functional particles for analysis and detection in vivo. Deng et al. have prepared a magnetic resonance probe by encapsulating photosensitizer Ce_6 and Fe_3O_4 nanoparticles in $mPEG_{2000}$-TK-C_{16} micelles (Deng et al., 2020). Under light irradiation, the compound cracks and the Fe_3O_4 particles are released to enhance the T_2-weighted magnetic resonance imaging signal. According to the change of T_2 signal, the detection of singlet oxygen in vivo can be achieved.

RNA-loaded magnetic liposome particles have been designed as biomarkers for treatment (Grippin et al., 2019). The iron oxide not only acts as a T_2- weighted contrast agent, but also promotes the activation and transfection of dendritic cells. Hsieh et al. have constructed MRI-based neurotransmitter responsive sensors composed of superparamagnetic nanoparticles conjugated to neurotransmitter analogs and engineered neurotransmitter binding proteins (Hsieh et al., 2019). In the presence of neurotransmitter analytes, the magnetic nanoclusters are reversible destroyed, thus changing the T_2- weighted signal, providing a promising way for the study of perturbation of neurochemical analytes.

Although MIONs as T_2- weighted contrast agent is the mainstream of current research, the study of T_1- weighted contrast agent is also proceeding with a series of breakthroughs. In 2009, Tromsdorf et al. have found that iron oxide smaller than 5 nm exhibits satisfactory T_1 relaxation performance for potential positive contrast agent (Tromsdorf et al., 2009). Under a field strength of 1.4 T, the r_1 value of the PEG-modified MION reaches 7.3 mM^{-1} s^{-1}, which is twice as the typical T_1 contrast agent (Magnevist). It has been revealed that ultra-small size is critical to apply MIONs as T_1 contrast agents (Tromsdorf et al., 2009), so subsequent studies have focused on the synthesis of uniform and ultra-small MIONs. In 2011, Kim et al. have used oleyl alcohol to reduce the thermal decomposition reaction temperature of iron oleate complex, and synthesized ultra-small iron oxide nanoparticles (1.5–3.7 nm) (Kim et al., 2011). The MIONs with particle sizes of 2.2 nm and 3.0 nm exhibit high r_1 relaxivity of 4.78 mM^{-1} s^{-1} and 4.77 mM^{-1} s^{-1} as well as low r_2/r_1 ratio of 3.67 and 6.12, respectively. Li et al. have used high-temperature coprecipitation to fabricate small MIONs with a diameter of 3.3 ± 0.5 nm and used poly(acrylic acid), poly(methacrylic acid) and their derives as versatile stabilizers to prevent nanoparticle aggregation (Li et al., 2012). The r_1 relaxivity is up to 8.3 mM^{-1} s^{-1} compared with 4.8 mM^{-1} s^{-1} for the commercial contrast agent Gd-DTPA. In vivo results also demonstrate enhanced tissue signals from T_1- weighted MRI. In particular, the liver signal is enhanced (~26%), much higher than that from the Gd-DTPA (<10%) under the same dosage.

15.1.3 Magnetic Particle Imaging

Magnetic particle imaging (MPI) is based on direct detection of the magnetic particles instead of the resonance signal (Gleich and Weizenecker, 2005). Compared to MRI, MPI possesses large penetration depth, positive contrast and almost no background interference generated by tissue (Yu et al., 2017). At present, MPI is usually based on iron oxide nanoparticles since they have clinically been proved to be nontoxic (Lemaster et al., 2018; Zheng et al., 2015; 2017). Song et al. have prepared Janus Fe_3O_4@ polymer nanoparticles, the MPI signal of which reaches three times that of the commercial MPI tracer (VivoTrax) and seven times that of the MRI contrast agent (Feraheme) with the same content of the Fe (Song et al., 2018). Recently, FeCo nanoparticles have also been used as an MPI tracer with much higher magnetic saturation (215 emu/g) than the Fe_3O_4 (21–80 emu/g) (Song et al., 2020). The FeCo particles coated carbon on the surface exhibit MPI signals 3.5 times greater than that of the VivoTrax.

15.1.4 Magnetic Hyperthermia Therapy

Magnetic hyperthermia therapy (MHT) is an advanced anti-tumor physical therapy with subtle side effects. It selectively kills tumor cells by introduction of MNPs at the target site and utilizes their thermal effect under alternating magnetic field. To improve the magnetic-thermal conversion efficiency (measured by specific absorption ratio/specific power loss, SAR/SPL), main strategies include adjusting the composition and surface modification of the MNPs (Du et al., 2019).

Pan et al. have utilized interfacial exchange interactions between hard and soft magnets to control the magnetocrystalline anisotropy for optimized SPL (Pan et al., 2020). $CoFe_2O_4$@$MnFe_2O_4$ nanoparticles have been synthesized and modified by 2,3-dimercaptosuccinic acid (DMSA) (Figure 15.1a). When the concentration of the nanoparticles is 10 mg/ml, the temperature could increase to nearly 80 °C in 180 s (Figure 15.1b), and the SLP value raises with the magnetic field intensity (Figure 15.1c).

For surface modifications, enhanced SAR has been reported with decreased thickness of surface coating due to the increase of Brownian-Néel relaxation and thermal conductivity (Liu et al., 2012). Zhang et al. have used caffeic acid instead of oleic acid to modify the surface of cobalt ferrite nanoparticles to form ultrathin coatings (<1 nm), giving rise to rapid increment of the tumor temperature to 42 °C in 120 s with the field of 27 kA/m and 115 kHz (Zhang et al., 2020).

Recently, MHT has also been combined with other methods such as PTT and chemotherapy to achieve optimized effect. For instance, $\gamma\text{-Fe}_2O_3$ has been combined with the CuS with high near infrared (NIR) absorption coefficient to produce cumulative heating (Curcio et al., 2019). Das et al. have prepared Ag@Fe_3O_4 nanoflowers as shown in Figure 15.2a–c, and found that compared with using the magnetocaloric therapy or

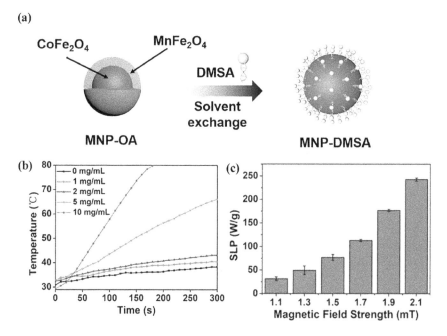

FIGURE 15.1 (a) Schematic synthesis of the $CoFe_2O_4$@$MnFe_2O_4$, (b) magnetic heating effect with varied $CoFe_2O_4$@$MnFe_2O_4$ concentration and (c) SLP values under different magnetic field strength. Reprinted with permission from Pan et al. (2020). Copyright (2020) American Chemical Society.

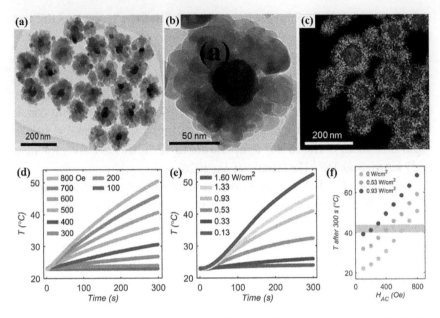

FIGURE 15.2 (a and b) TEM images of the Ag@Fe$_3$O$_4$ nanoflowers and (c) elemental mapping for the Fe and Ag in green and red color. Effects of (d) alternating field and (e) heat power on the change of temperature, and (f) their combined effects. Reprinted with permission from Das et al. (2016). Copyright (2016) American Chemical Society.

photothermal therapy alone (Figure 15.2d and e), much reduced combination of magnetic field intensity and power density is required to achieve the target temperature 40 °C as shown in Figure 15.2f (Das et al., 2016). Consequently, damage to the health tissues can be minimized while killing the target cancer cells.

MHT has also been combined with chemotherapy since the generated heat promotes the diffusion rate of drugs in vivo during synergistic chemomagnetic hyperthermia therapy. Yao et al. have synthesized composites composed of Fe$_3$O$_4$/SiO$_2$ mesoporous nanoparticles (MMSN) as carriers for the drug DOX and graphene quantum dots as caps to prevent drug leakage (Yao et al., 2017). While chemotherapy can only reduce the viability of cancer cells to ~60%, the chemo-magnetic hyperthermia therapy gives rise to significantly decreased viability of ~5%.

15.2 MAGNETIC MATERIALS FOR CATALYSIS

Due to the unique magnetic response and behavior under applied field, magnetic nanoparticles have also been used in catalysis (Chávez et al., 2020; Li et al., 2020; Wang et al., 2020). The MNPs can either act as the support for various catalysts or the catalyst itself. The effects of magnetic materials on catalysis can be summarized as magnetic separation for catalyst recycling and direct or indirect involvement in the reaction pathways.

15.2.1 Magnetic Separation for Catalyst Recycling

Most of the catalysts are nanoparticles, making their recycling difficult after the reaction (Qin et al., 2020; Shylesh et al. 2010). A possible solution is to use the magnetic separation by incorporation a magnetic component into the catalytic system. Core-shell Fe_3O_4@CuMgAl-LDH nanocomposite has been prepared via coprecipitation as the catalyst for the phenol hydroxylation, which could be effectively separated from the reaction solution by applying an external magnetic field of 1500 G (Zhang et al., 2013). Although the composition and the core-shell structure of recycled Fe_3O_4@CuMgAl-LDH is well maintained, the phenol conversion efficiency decreases from 47.2% of the fresh catalyst to 39.1% and 38.3% for the catalyst recycled once and twice. Fe_3O_4-chitosan nanoparticles have been used as the catalyst during the synthesis of 2-amino-4H-chromenes (Gao et al., 2020). Not only could the catalyst be easily separated by an external magnet with neglectable mass loss, but the catalytic activity has also been maintained after recycling for four times. Similar separation effect of magnetic catalysts has also observed in various systems such as Fe_2O_3@CuMgAl (Xia et al., 2014), Fe_3O_4-CuO@meso-SiO_2 (Zhang et al., 2014) and chitosan-encapsulated Fe_3O_4/SiO_2-NH_2 (Veisi et al., 2020).

15.2.2 Direct Involvement of Magnetic Materials in the Catalytic Process

Only a few studies reveal direct catalytic effects of magnetic materials, mainly focusing on the oxygen evolution reaction (OER) and hydrogen evolution reaction (HER) in electrocatalysis. The OER takes place at the anode where the H_2O molecule is oxidized to generate O_2 (Chai et al., 2017). The OER is the key factor that limits the efficiency of the overall water splitting, since it is a slow four-electron process and requires much energy to proceed (Gracia 2017; Gracia et al. 2018). The formation of O–O bond requires spin conservation to generate oxygen molecule of paramagnetic triplet state. Since spin polarization on the surface of catalyst could facilitate the parallel spin alignment of the oxygen atoms, it provides a possible way to enhance the OER efficiency (Mtangi et al., 2015; 2017). Local field induced by magnetic catalyst can potentially promote such spin alignment. Garcés-Pineda et al. have fabricated multiple OER catalysts with different magnetic features and examined their electrocatalytic performance with and without the introduction of magnetic field (Garcés-Pineda et al., 2019). Magnetocurrent could be obtained for all the samples with distinctive enhancement in the current density for the magnetic catalysts (Figure 15.3a), exhibiting linear relationship with the magnetization (Figure 15.3b). Maximum current density enhancement from 24 mA cm^{-2} to 40 mA cm^{-2} at 1.65 V has been achieved for the NiZnFe$_4$O$_x$ with the largest magnetization, promoting parallel alignment of the oxygen to form the O–O bond for improved OER efficiency and the overall water splitting.

Magnetic materials also play a positive role during the HER process by facilitating the migration of electrons. Zhou et al. have fabricated ferromagnetic bowl-like MoS$_2$ flakes with asymmetric up and down spin bands (Figures 15.3c and d) (Zhou et al., 2020). Enhanced HER has been achieved under vertical magnetic field with almost doubled current density. Compared to the nonmagnetic bilayer MoS$_2$, it is possible for the electrons of the ferromagnetic MoS$_2$ to take extra energy from the external magnetic field (Figure 15.3e) for facilitated migration from the conductive glass carbon electrode to the active sites (Figure 15.3f).

15.2.3 Indirect Involvement of Magnetic Materials in the Catalytic Process

Applied external magnetic field and/or local field induced by the magnetic catalysts may affect the migration of the surrounding magnetic reactants or charged particles by Lorentz force. For instance, magnetic field has been reported to influence the movement of paramagnetic species such as O$_2$ for regulated ORR process (Wu et al., 2016). Monzon et al. have fabricated substrates containing hemispherical Fe, Co and Zn metallic crystals (Figure 15.4a) and investigated the magnetic field effect generated by these particles on the ORR (Monzon et al., 2012). Under an applied magnetic field, increased current of 3.0 ± 0.7%, 7.8 ± 1.2% and 11.6 ±1.8% has been observed for the samples containing Zn, Co and Fe, respectively.

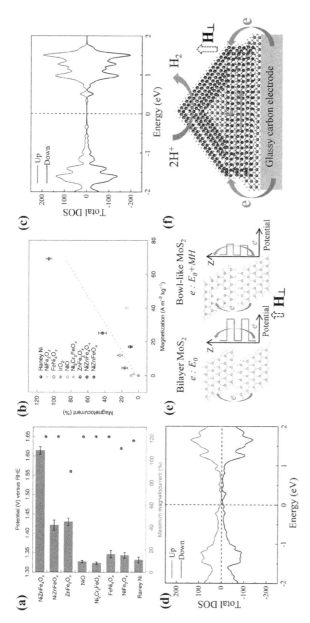

FIGURE 15.3 (a) Maximum magnetocurrent of various catalysts and (b) its correlation with the magnetization. Reprinted with permission from (Garcés-Pineda et al., 2019). Copyright (2019) Springer Nature. Total density of states of the (c) bilayer and (d) bowl-like MoS_2. (e) Electron transfer in the bilayer and bowl-like MoS_2 with applied magnetic field, and (f) that in the bowl-like MoS_2 flakes during HER. Reprinted with permission from (Zhou et al., 2020). Copyright (2020) American Chemical Society.

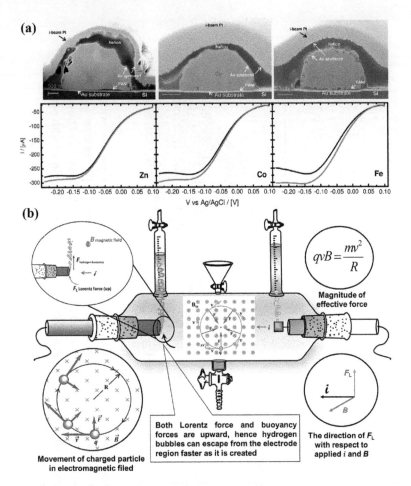

FIGURE 15.4 (a) Morphology of substrates modified with micro-hemispheres and the cathodic scans for the ORR with/without magnetic field in red and black color, respectively.Reprinted with permission from (Monzon et al., 2012). Copyright (2012) American Chemical Society. (https://pubs.acs.org/doi/10.1021/cm301766s, and further permissions related to the material excerpted should be directed to the ACS) (b) Mechanism of enhancement in the HER with applied magnetic field. Reprinted with permission from (Elias and Hegde, 2017). Copyright (2017) Springer Nature.

Such improvement has been attributed to the Lorentz force, which causes the convection of paramagnetic reactant O_2 in the vicinity of the electrode surface and increases their flux toward the electrode. Since the Co and Fe are soft magnetic materials, they induce strong local fields for more significant improvement of the current density. Similar enhancement via Lorentz force has also observed in the HER process of water electrolysis using Ni-W alloy as the electrode (Elias and Hegde, 2017). The improvement of HER efficiency has been attributed to the magnetohydrodynamic force-induced convection arising from the Lorentz force and the effective disentanglement of H_2 from the electrode (Figure 15.4b).

Under a high-frequency, alternating external field, the magnetic heating effect takes place for the magnetic materials, promoting catalytic reaction on their surfaces. Niether et al. have synthesized FeC@Ni nanoparticles and used it as the electrocatalyst for water splitting (Figure 15.5a) (Niether et al., 2018). The magnetic heating on the surface of FeC-Ni nanoparticles effectively decreases the overpotential of the OER and HER by ~200 mV

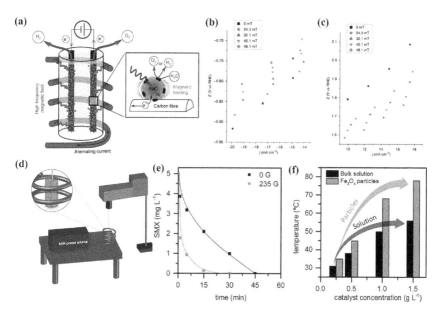

FIGURE 15.5 (a) Illustration showing the application of alternating magnetic field with the reactor operating inside the coils. Quasistationary potential values of the (b) HER and (c) OER under varied field strength. Reprinted with permission from Niether et al. (2018). Copyright (2018) Springer Nature. (d) Magnetic induction heating for the CWPO experiments. (e) Concentration of the sulfamethoxazole (SMX) as a function of time with/without the magnetic field, and (f) temperature of the bulk solution compared with that of the catalyst particles under magnetic induction heating. Reprinted with permission from Munoz et al. (2020). Copyright (2020) Elsevier.

and 100 mV, respectively as shown in Figure 15.5b and c. Usually, such improvement of OER requires the reaction to operate at ~200 °C, while the temperature of the whole reaction cell only increases by 3.5 °C, demonstrating effective local heating induction of the magnetic particles under the alternating field. A similar promotion effect has also observed in the catalytic wet peroxide oxidation (CWPO) for wastewater treatment. Munoz et al. have used magnetite mineral Fe_3O_4 as the catalyst and investigated its magnetic induction effect under alternating field as shown in Figure 15.5d (Munoz et al., 2020). The introduction of an alternating field improves the oxidation rate by three times for significantly enhanced CWPO efficiency. Contrast experiments have been carried out at the same temperature of the bulk solution using a conventional heating plate rather than magnetic induction. As shown in Figure 15.5e, the CWPO oxidation process is much faster using magnetic heating than the conventional heating due to the higher temperature of Fe_3O_4 particles than the bulk solution (Figure 15.5f).

15.3 MAGNETIC MATERIALS FOR OTHER AREAS

15.3.1 Micro-Magnetic Robots

The development of micro-robot is technologically important for unmanned aerial vehicles, exploration of unknown environment, minimally invasive surgery and targeted drug delivery. Magnetic-field-driven micro-robots have attracted much attention due to their controllability, for which appropriate use of magnetic materials is the key. Su et al. have fabricated tadpole-like polycaprolactone/Fe_3O_4 micro-robots, for which two motion modes, including the rolling mode and the propulsion mode have been achieved under external magnetic field as illustrated in Figure 15.6 (Su et al., 2021). Under the rolling motion, the micro-robots can reach the destination at a speed of ~2 mm s^{-1}, while the propulsion motion (0–340 μm s^{-1}) allows handling of a micro-cargo.

In another work, Kim et al. have managed to program ferromagnetic domains by controllable 3D printing with applied magnetic field to the dispensing nozzle (Figure 15.7) (Kim et al., 2018). NdFeB magnetic microparticles have been used as the printing material together with silica nanoparticles and silicon rubber. The NdFeB particles can be magnetized with different directions by adjusting the magnetic field applied near the nozzle for delicate design of the artificial magnetic domains. The assembly of regions with varied magnetization directions give rise to robots with

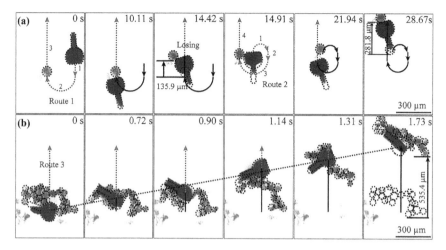

FIGURE 15.6 Optical microscopic images showing the (a) propulsion mode and (b) rolling mode of the micro-robot under an external field of 1.85 mT and 4 Hz (Su et al., 2021). (Open access article)

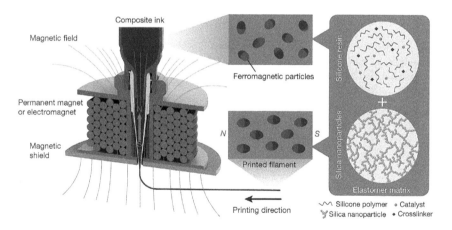

FIGURE 15.7 Schematic illustration showing the 3D printing where composite ink containing ferromagnetic particles, silicone resin and silica nanoparticles is used, among which the ferromagnetic particles can be reoriented by the magnetic field applied around the dispensing nozzle. Reprinted with permission from Kim et al. (2018). Copyright (2018) Springer Nature.

multiple functions under a controlling external magnetic field. Derived from complex shape changes, the robots could grasp a rapidly moving object, pack a drug and deliver to a designated location and released, and jump with a speed up to 250 mm/s.

In order to integrate multiple deformation characteristics, micron-sized Fe_3O_4 and NdFeB particles have been embedded in a polyacrylate-based shape memory polymer matrix to form a magnetic shape memory polymer composite (Ze et al., 2020). The soft magnetic Fe_3O_4 generates inductive heating to soften the material while the hard magnetic NdFeB allows shape change under applied magnetic field. Consequently, the composite exhibit adjustable stiffness for simultaneous low-temperature shape memory and high-temperature fast drive.

15.3.2 Magnetic Fluids and Magnetic Fluidic Platform

In microfluidic devices with miniaturized fluid channels, solid walls of the fluid channels are critical for limited flow rate at a given pressure. Dunne et al. have adopted the magnetic fluid technology to break the limit of solid wall with a wall-less liquid channel surrounded by an immiscible magnetic fluid which can be stabilized by a quadrupolar magnetic field (Dunne et al., 2020). As such zero magnetic field along the centerline can be achieved to allow the wall-free magnetic constraint (Figure 15.8).

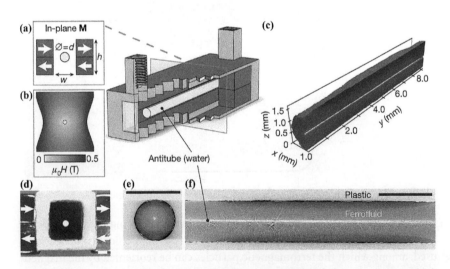

FIGURE 15.8 (a) In-plane quadrupolarly arranged permanent magnets (red and blue) and (b) contour plot of the magnetic field, where water (yellow) is stabilized at the center. (c) Synchrotron x-ray tomographic image of the ferrofluid (blue) confined water antitube (yellow) as well as its (d) optical end-, (e) x-ray end- and (f) x-ray side-views. Reprinted with permission from Dunne et al. (2020). Copyright (2020) Springer Nature.

The liquid-in-liquid design is mainly through the injection of ferrofluid into a water-filled quadrupole channel, and the antitube (yellow) are surrounded by the ferrofluid (blue) as shown in Figure 15.8c. The diameter of the antitube depends on the susceptibility of the ferrofluid and the interface energy between the antitube and the ferrofluid.

Guo et al. have designed a magnetic digital microfluidics platform (MDMP) using magnetic probe to accurately and reversibly manipulate liquid droplets (Guo et al., 2019). The platform consists of three main components, including a magnetic responsive layer (MRL) with micro-Fe particles dispersed in polydimethylsiloxane (PDMS), an actuated feedback layer containing low cross-linked PDMS, and a slippery surface with hollow nano-Si coating infused oil (Figure 15.9a). The applied magnetic field attracts the MRL which deforms the actuated feedback layer reversibly with the assistance of the elastic PDMS (Figure 15.9b) and provides a geopotential gradient on the surface to drive the droplet. The MDMP system is capable of operating a wide range of liquids in a variety of environments, exhibiting great potential in the fields such as biological testing and chemical reactions.

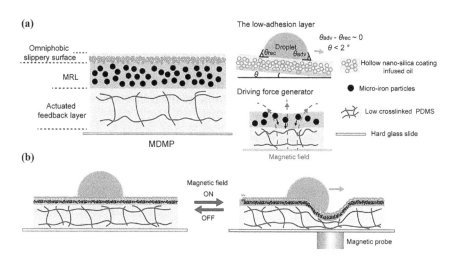

FIGURE 15.9 (a) Schematic of the magnetic digital microfluidics platform and (b) mechanism for controllable droplet movement with/without the application of the external magnetic field. Reprinted with permission from Guo et al. (2019). Copyright (2019) John Wiley and Sons.

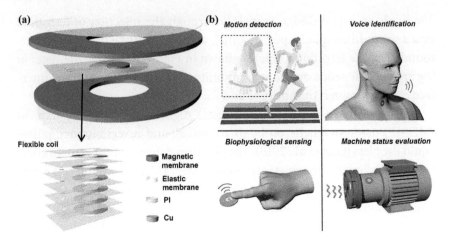

FIGURE 15.10 (a) Schematic diagram showing the setup of the sensor and (b) its applications in various areas.Reprinted with permission from Zhao et al. (2020). Copyright (2020) John Wiley and Sons.

15.3.3 Magneto-Electric Vibration Sensor

The requirement of sensing and energy collection in different environments leads to the development of flexible piezoelectric and triboelectric components. Complete flexible electromagnetic devices usually require finite and complex magnetic fields, which are difficult to achieve even using well-arranged hard magnets. Zhao et al. have developed a flexible micro-electro-mechanical system (MEMS) consisting of multi-layer flexible coils sandwiched between two pairs of magnetic and elastic membranes (Figure 15.10a) (Zhao et al., 2020). External force and vibration generate changes in the field distribution of the magnetic membranes, and induces electromotive force in the coils. Such sensor is capable to provide frequency response range between 1 Hz and 10 kHz. It can also be flexibly installed on soft and curved surface for applications such as motion detection, voice identification, health monitoring and machine condition assessment (Figure 15.10b).

15.4 SUMMARY AND PERSPECTIVES

Magnetic materials, due to their responses to externally applied magnetic fields, such as motion, relaxation and induction, have been widely explored in areas including and beyond biomedicine and catalysis. Combination of the special function of magnetic materials not only brings synergistic effects in various fields, but also provides inspirations for stand-alone applications. It can be expected that in the near future, the increasing trend to apply magnetic materials interdisciplinary will further develop along the directions of (i) delicate design and fabrication of the materials or devices incorporating magnetic components; (ii) manipulation of the strength or frequency for the external field in the generation of varied magnetic responses; and (iii) bold and creative choice of the application areas.

REFERENCES

Chai, G. L., K. P. Qiu, M. Qiao, M. M. Titirici, C. X. Shang, and Z. X. Guo. 2017. Active sites engineering leads to exceptional ORR and OER bifunctionality in P, N Co-doped graphene frameworks. *Energy & Environmental Science* 10:1186–95.

Chávez, A. M., D. H. Quiñones, A. Rey, F. J. Beltrán, and P. M. Álvarez. 2020. Simulated solar photocatalytic ozonation of contaminants of emerging concern and effluent organic matter in secondary effluents by a reusable magnetic catalyst. *Chemical Engineering Journal* 398:125642.

Curcio, A., A. K. A. Silva, S. Cabana, A. Espinosa, B. Baptiste, N. Menguy, C. Wilhelm, and A. Abou-Hassan. 2019. Iron oxide nanoflowers@CuS hybrids for cancer tri-therapy: interplay of photothermal therapy, magnetic hyperthermia and photodynamic therapy. *Theranostics* 9:1288–302.

Das, R., N. Rinaldi-Montes, J. Alonso, Z. Amghouz, E. Garaio, J. A. Garcia, P. Gorria, J. A. Blanco, M. H. Phan, and H. Srikanth. 2016. Boosted hyperthermia therapy by combined AC magnetic and photothermal exposures in Ag/Fe$_3$O$_4$ nanoflowers. *ACS Applied Materials & Interfaces* 8:25162–9.

Deng, K., B. Wu, C. X. Wang, Q. Wang, H. Yu, J. M. Li, K. H. Li, H. Y. Zhao, and S. W. Huang. 2020. An oxidation-enhanced magnetic resonance imaging probe for visual and specific detection of singlet oxygen generated in photodynamic cancer therapy in vivo. *Advanced Healthcare Materials* 9:2000533.

Du, Y., X. Liu, Q. Liang, X. J. Liang, and J. Tian. 2019. Optimization and design of magnetic ferrite nanoparticles with uniform tumor distribution for highly sensitive MRI/MPI performance and improved magnetic hyperthermia therapy. *Nano Letters* 19:3618–26.

Dunne, P., T. Adachi, A. A. Dev, A. Sorrenti, L. Giacchetti, A. Bonnin, C. Bourdon, P. H. Mangin, J. M. D. Coey, B. Doudin, and T. M. Hermans. 2020. Liquid flow and control without solid walls. *Nature* 581:58–62.

Edward, M., J. A. Quinn, S. Mukherjee, M. Bv Jensen, A. G. Jardine, P. B. Mark, and A. D. Burden. 2008. Gadodiamide contrast agent 'activates' fibroblasts: a possible cause of nephrogenic systemic fibrosis. *Journal of pathology* 214:584–93.

Elias, L., and A. C. Hegde. 2017. Effect of magnetic field on HER of water electrolysis on Ni-W alloy. *Electrocatalysis* 8:375–82.

Feng, L., R. Xie, C. Wang, S. Gai, F. He, D. Yang, P. Yang, and J. Lin. 2018. Magnetic targeting, tumor microenvironment-responsive intelligent nanocatalysts for enhanced tumor ablation. *ACS Nano* 12:11000–12.

Fraum, T. J., D. R. Ludwig, M. R. Bashir, and K. J. Fowler. 2017. Gadolinium-based contrast agents: a comprehensive risk assessment. *Journal of Magnetic Resonance Imaging* 46:338–53.

Gao, G., J. Q. Di, H. Y. Zhang, L. P. Mo, and Z. H. Zhang. 2020. A magnetic metal organic framework material as a highly efficient and recyclable catalyst for synthesis of cyclohexenone derivatives. *Journal of Catalysis* 387:39–46.

Garcés-Pineda, F. A., M. Blasco-Ahicart, D. Nieto-Castro, N. López, and J. R. Galán-Mascarós. 2019. Direct magnetic enhancement of electrocatalytic water oxidation in alkaline media. *Nature Energy* 4:519–25.

Gleich, B., and J. Weizenecker. 2005. Tomographic imaging using the nonlinear response of magnetic particles. *Nature* 435:1214–7.

Gracia, J. 2017. Spin dependent interactions catalyse the oxygen electrochemistry. *Physical Chemistry Chemical Physics* 19:20451–6.

Gracia, J., R. Sharpe, and J. Munarriz. 2018. Principles determining the activity of magnetic oxides for electron transfer reactions. *Journal of Catalysis* 361:331–8.

Grippin, A. J., B. Wummer, T. Wildes, K. Dyson, V. Trivedi, C. Yang, M. Sebastian, H. R. Mendez-Gomez, S. Padala, M. Grubb, M. Fillingim, A. Monsalve, E. J. Sayour, J. Dobson, and D. A. Mitchell. 2019. Dendritic cell-activating magnetic nanoparticles enable early prediction of antitumor response with magnetic resonance imaging. *ACS Nano* 13:13884–98.

Guo, J. C., D. H. Wang, Q. Q. Sun, L. X. Li, H. X. Zhao, D. S. Wang, J. X. Cui, L. Q. Chen, and X. Deng. 2019. Omni-liquid droplet manipulation platform. *Advanced Materials Interfaces* 6:1–9.

Hsieh, V., S. Okada, H. Wei, I. Garcia-Alvarez, A. Barandov, S. R. Alvarado, R. Ohlendorf, J. Fan, A. Ortega, and A. Jasanoff. 2019. Neurotransmitter-responsive nanosensors for T_2-weighted magnetic resonance imaging. *Journal of American Chemical Society* 141:15751–4.

Kim, B. H., N. Lee, H. Kim, K. An, Y. I. Park, Y. Choi, K. Shin, Y. Lee, S. G. Kwon, H. B. Na, J. G. Park, T. Y. Ahn, Y. W. Kim, W. K. Moon, S. H. Choi, and T. Hyeon. 2011. Large-scale synthesis of uniform and extremely small-sized iron oxide nanoparticles for high-resolution T_1 magnetic resonance Imaging contrast agents. *Journal of American Chemical Society* 133:12624–31.

Kim, Y., H. Yuk, R. Zhao, S. A. Chester, and X. H. Zhao. 2018. Printing ferromagnetic domains for untethered fast-transforming soft materials. *Nature* 558:274–9.

Lemaster, J. E., F. Chen, T. Kim, A. Hariri, and J. V. Jokerst. 2018. Development of a trimodal contrast agent for acoustic and magnetic particle imaging of stem cells. *ACS Applied Nano Materials* 1:1321–31.

Li, Y. Y., A. Chatterjee, L. B. Chen, F. L. Y. Lam, and X. J. Hu. 2020. Pd doped Co functionalized SBA-15 as an active magnetic catalyst for low temperature solventless additive-base-free selective oxidation of benzyl alcohol. *Molecular Catalysis* 488:110869.

Li, Z., P. W. Yi, Q. Sun, H. Lei, Z. H. Li, Z. H. Zhu, S. C. Smith, M. B. Lan, and G. Q. Lu. 2012. Ultrasmall water-soluble and biocompatible magnetic iron oxide nanoparticles as positive and negative dual contrast agents. *Advanced Functional Materials* 22:2387–93.

Liang, P. C., Y. C. Chen, C. F. Chiang, L. R. Mo, S. Y. Wei, W. Y. Hsieh, and W. L. Lin. 2016. Doxorubicin-modified magnetic nanoparticles as a drug delivery system for magnetic resonance imaging-monitoring magnet-enhancing tumor chemotherapy. *International Journal of Nanomedicine* 11:2021–37.

Liu, J. F., Z. Lan, C. Ferrari, J. M. Stein, E. Higbee-Dempsey, L. Yan, A. Amirshaghaghi, Z. Cheng, D. Issadore, and A. Tsourkas. 2020. Use of oppositely polarized external magnets to improve the accumulation and penetration of magnetic nanocarriers into solid tumors. *ACS Nano* 14:142–52.

Liu, X. L., H. M. Fan, J. B. Yi, Y. Yang, E. S. G. Choo, J. M. Xue, D. D. Fan, and J. Ding. 2012. Optimization of surface coating on Fe_3O_4 nanoparticles for high performance magnetic hyperthermia agents. *Journal of Materials Chemistry* 22:8235–44.

Monzon, L. M. A., K. Rode, M. Venkatesan, and J. M. D. Coey. 2012. Electrosynthesis of iron, cobalt, and zinc microcrystals and magnetic enhancement of the oxygen reduction reaction. *Journal of Chemistry of Materials* 24:3878–85.

Mtangi, W., V. Kiran, C. Fontanesi, and R. Naaman. 2015. Role of the electron spin polarization in water splitting. *Journal of Physical Chemistry Letters* 6:4916–22.

Mtangi, W., F. Tassinari, K. Vankayala, A. Vargas-Jentzsch, B. Adelizzi, A. R. A. Palmans, C. Fontanesi, E. W. Meijer, and R. Naaman. 2017. Control of electrons' spin eliminates hydrogen peroxide formation during water splitting. *Journal of American Chemical Society* 139:2794–8.

Munoz, M., J. Nieto-Sandoval, E. Serrano, Z. M. de Pedro, and J. A. Casas. 2020. CWPO intensification by induction heating using magnetite as catalyst. *Journal of Environmental Chemical Engineering* 8:104085.

Niether, C., S. Faure, A. Bordet, J. Deseure, M. Chatenet, J. Carrey, B. Chaudret, and A. Rouet. 2018. Improved water electrolysis using magnetic heating of FeC-Ni core-shell nanoparticles. *Nature Energy* 3:476–83.

Pan, J., P. Hu, Y. Guo, J. Hao, D. Ni, Y. Xu, Q. Bao, H. Yao, C. Wei, Q. Wu, and J. Shi. 2020. Combined magnetic hyperthermia and immune therapy for primary and metastatic tumor treatments. *ACS Nano* 14:1033–44.

Qin, L., R. Ru, J. W. Mao, Q. S. Meng, Z. Fan, X. Li, and G. L. Zhang. 2020. Assembly of MOFs/polymer hydrogel derived Fe_3O_4-CuO@ hollow carbon spheres for photochemical oxidation: freezing replacement for structural adjustment. *Applied Catalysis B: Environmental* 269:118754.

Shylesh, S., V. Schuenemann, and W. R. Thiel. 2010. Magnetically separable nanocatalysts: bridges between homogeneous and heterogeneous catalysis. *Angewandte Chemie International Edition* 49:3428–59.

Song, G., M. Chen, Y. Zhang, L. Cui, H. Qu, X. Zheng, M. Wintermark, Z. Liu, and J. Rao. 2018. Janus iron oxides@semiconducting polymer nanoparticle tracer for cell tracking by magnetic particle imaging. *Nano Letters* 18:182–9.

Song, G., M. Kenney, Y. S. Chen, X. Zheng, Y. Deng, Z. Chen, S. X. Wang, S. S. Gambhir, H. Dai, and J. Rao. 2020. Carbon-coated FeCo nanoparticles as sensitive magnetic-particle-imaging tracers with photothermal and magnetothermal properties. *Nature Biomedical Engineering* 4:325–34.

Su, Y. C., T. Qiu, W. Song, X. J. Han, M. M. Sun, Z. Wang, H. Xie, M. D. Dong, and M. l. Chen. 2021. Melt electrospinning writing of magnetic microrobots. *Advanced Science* 8:1–7.

Tromsdorf, U. I., O. T. Bruns, S. C. Salmen, U. Beisiegel, and H. Weller. 2009. A highly effective, nontoxic T-1 MR contrast agent based on ultrasmall pegylated iron oxide nanoparticles. *Nano Letters* 9:4434–40.

Veisi, H., T. Ozturk, B. Karmakar, T. Tamoradi, and S. Hemmati. 2020. In situ decorated Pd NPs on chitosan-encapsulated Fe_3O_4/SiO_2-NH_2 as magnetic catalyst in Suzuki-Miyaura coupling and 4-nitrophenol reduction. *Carbohydrate Polymers* 235:115966.

Wang, W. Z., Z. W. Dai, R. Jiang, Q. Li, X. Zheng, W. Liu, Z. G. Luo, Z. M. Xu, and J. Peng. 2020. Highly phosphatized magnetic catalyst with electron transfer induced by quaternary synergy for efficient dehydrogenation of ammonia borane. *ACS Applied Materials & Interfaces* 12:43854–63.

Wu, Y., A. Bhalla, and R. Y. Guo. 2016. Magnetic field tunable capacitive dielectric: ionic-liquid sandwich composites. *Materials Research Express* 3:036102.

Xia, S. X., W. C. Du, L. P. Zheng, P. Chen, and Z. Y. Hou. 2014. A thermally stable and easily recycled core-shell Fe_2O_3@CuMgAl catalyst for hydrogenolysis of glycerol. *Catalysis Science & Technology* 4:912–6.

Xuan, M., J. Shao, J. Zhao, Q. Li, L. Dai, and J. Li. 2018. Magnetic mesoporous silica nanoparticles cloaked by red blood cell membranes: applications in cancer therapy. *Angewandte Chemie International Edition* 57:6049–53.

Yao, X., X. Niu, K. Ma, P. Huang, J. Grothe, S. Kaskel, and Y. Zhu. 2017. Graphene quantum dots-capped magnetic mesoporous silica nanoparticles as a multifunctional platform for controlled drug delivery, magnetic hyperthermia, and photothermal therapy. *Small* 13:1602225.

Yu, E. Y., M. Bishop, B. Zheng, R. M. Ferguson, A. P. Khandhar, S. J. Kemp, K. M. Krishnan, P. W. Goodwill, and S. M. Conolly. 2017. Magnetic particle imaging: a novel in vivo imaging platform for cancer detection. *Nano Letters* 17:1648–54.

Yu, J., W. Yin, X. Zheng, G. Tian, X. Zhang, T. Bao, X. Dong, Z. Wang, Z. Gu, X. Ma, and Y. Zhao. 2015. Smart MoS_2/Fe_3O_4 nanotheranostic for magnetically targeted photothermal therapy guided by magnetic resonance/photoacoustic imaging. *Theranostics* 5:931–45.

Ze, Q. J., X. Kuang, S. Wu, J. Wong, S. M. Montgomery, R. D. Zhang, J. M. Kovitz, F. Y. Yang, H. J. Qi, and R. K. Zhao. 2020. Magnetic shape memory polymers with integrated multifunctional shape manipulation. *Advanced Materials* 32:1–8.

Zhang, H., G. Y. Zhang, X. Bi, and X. T. Chen. 2013. Facile assembly of a hierarchical core@ shell Fe_3O_4@ CuMgAl-LDH (layered double hydroxide) magnetic nanocatalyst for the hydroxylation of phenol. *Journal of Materials Chemistry A* 1:5934–42.

Zhang, L., Z. Liu, Y. Liu, Y. Wang, P. Tang, Y. Wu, H. Huang, Z. Gan, J. Liu, and D. Wu. 2020. Ultrathin surface coated water-soluble cobalt ferrite nanoparticles with high magnetic heating efficiency and rapid in vivo clearance. *Biomaterials* 230:119655.

Zhang, X. W., G. Wang, M. Yang, Y. Luan, W. J. Dong, R. Dang, H. Y. Gao, and J. Yu. 2014. Synthesis of a Fe_3O_4-CuO@meso-SiO_2 nanostructure as a magnetically recyclable and efficient catalyst for styrene epoxidation. *Catalysis Science & Technology* 4:3082–9.

Zhao, Y. C., S. H. Gao, X. Zhang, W. X. Huo, H. Xu, C. Chen, J. Li, K. X. Xu, and X. Huang. 2020. Fully flexible electromagnetic vibration sensors with annular field confinement origami magnetic membranes. *Advanced Functional Materials* 30:1–10.

Zheng, B., T. Vazin, P. W. Goodwill, A. Conway, A. Verma, E. U. Saritas, D. Schaffer, and S. M. Conolly. 2015. Magnetic particle imaging tracks the long-term fate of in vivo neural cell implants with high image contrast. *Scientific Reports* 5:14055.

Zheng, B., E. Yu, R. Orendorff, K. Lu, J. J. Konkle, Z. W. Tay, D. Hensley, X. Y. Zhou, P. Chandrasekharan, E. U. Saritas, P. W. Goodwill, J. D. Hazle, and S. M. Conolly. 2017. Seeing SPIOs directly in vivo with magnetic particle imaging. *Molecular Imaging and Biology* 19:385–90.

Zhou, W. D., M. Y. Chen, M. M. Guo, A. J. Hong, T. Yu, X. F. Luo, C. L. Yuan, W. Lei, and S. G. Wang. 2020. Magnetic enhancement for hydrogen evolution reaction on ferromagnetic MoS_2 catalyst. *Nano Letters* 20:2923–30.

FURTHER READING

Zhu, K., Y. Ju, J. Xu, Z. Yang, S. Gao, and Y. Hou. 2018. Magnetic nanomaterials: chemical design, synthesis, and potential applications. *Accounts of Chemical Research* 51: 404–13.

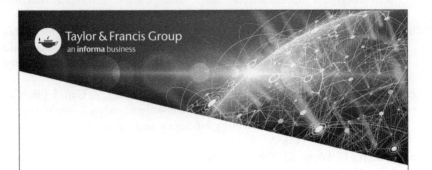

For Product Safety Concerns and Information please contact our EU
representative GPSR@taylorandfrancis.com
Taylor & Francis Verlag GmbH, Kaufingerstraße 24, 80331 München, Germany

www.ingramcontent.com/pod-product-compliance
Ingram Content Group UK Ltd.
Pitfield, Milton Keynes, MK11 3LW, UK
UKHW031503140425
457368UK00006B/21